Emergency Planning

Independent Study 235.a

May 2010

 FEMA

TABLE OF CONTENTS

Page

Course Overview..1

Unit 1: Course Introduction

Introduction 1.1
How To Take This Course 1.1
Case Study: Why Plan?..1.4
Course Goals 1.6
Goal Setting 1.6
Activity: Personal Learning Goals 1.7
Unit Summary 1.8
For More Information 1.8

Unit 2: The Planning Process

Introduction and Unit Overview..2.1
Mandates: Incident Management and Coordination Systems 2.1
What These Changes Mean To You...2.6
The Emergency Planning Process..2.7
Who Should Be Involved?..2.8
How To Get the Team Together 2.11
How Should the Team Operate? 2.12
Activity: Organizational Roles and Individual Skills....................................2.15
Unit Summary 2.16
For More Information 2.17
Knowledge Check 2.18

Unit 3: Threat Analysis

Introduction and Unit Overview..3.1
The Threat Analysis Process 3.1
Step 1: Identifying Threats 3.2
Step 2: Profiling Threats 3.3
Activity: Profiling a Threat 3.4
Step 3: Developing a Community Profile 3.6
Step 4: Determining Vulnerability..3.8
Activity: Prioritizing Risks 3.11
Step 5: Creating and Applying Scenarios 3.12
Activity: Threat Analysis...3.13
Unit Summary 3.17
For More Information 3.18
Knowledge Check 3.19

Unit 4: The Basic Plan

Introduction and Unit Overview..4.1
Components of a Basic Plan...4.1
Activity: Basic Plan Review...4.6
Activity: Reviewing Your Community's Basic Plan...4.9
Unit Summary 4.11
Knowledge Check 4.12

Unit 5: Annexes and Appendices

Introduction and Unit Overview...5.1
Annexes vs. Appendices—What Is the Difference? 5.1
Functional Annexes 5.2
Activity: Reviewing Your EOP's Functional Annexes..5.4
Hazard-, Threat-, and Incident-Specific Appendices 5.6
Annex and/or Appendix Implementing Instructions...5.8
Activity: Appendix Review...5.9
Unit Summary 5.10
Knowledge Check..5.11

Unit 6: Implementing Instructions

Introduction and Unit Overview...6.1
What Are Implementing Instructions? 6.1
Activity: Which Type Is Best?..6.6
Who Uses Implementing Instructions? 6.8
Activity: Identifying Possible Agency Implementing Instructions...............................6.9
Unit Summary 6.10
Knowledge Check..6.11

Unit 7: Course Summary

Introduction 7.1
The Planning Process...7.1
Threat Analysis 7.3
The Basic Plan..7.5
Annexes and Appendices 7.6
Implementing Instructions 7.7
Final Steps..7.7

Appendix A: Sample Plan: Jefferson County
Appendix B: Job Aids
Appendix C: Acronym List

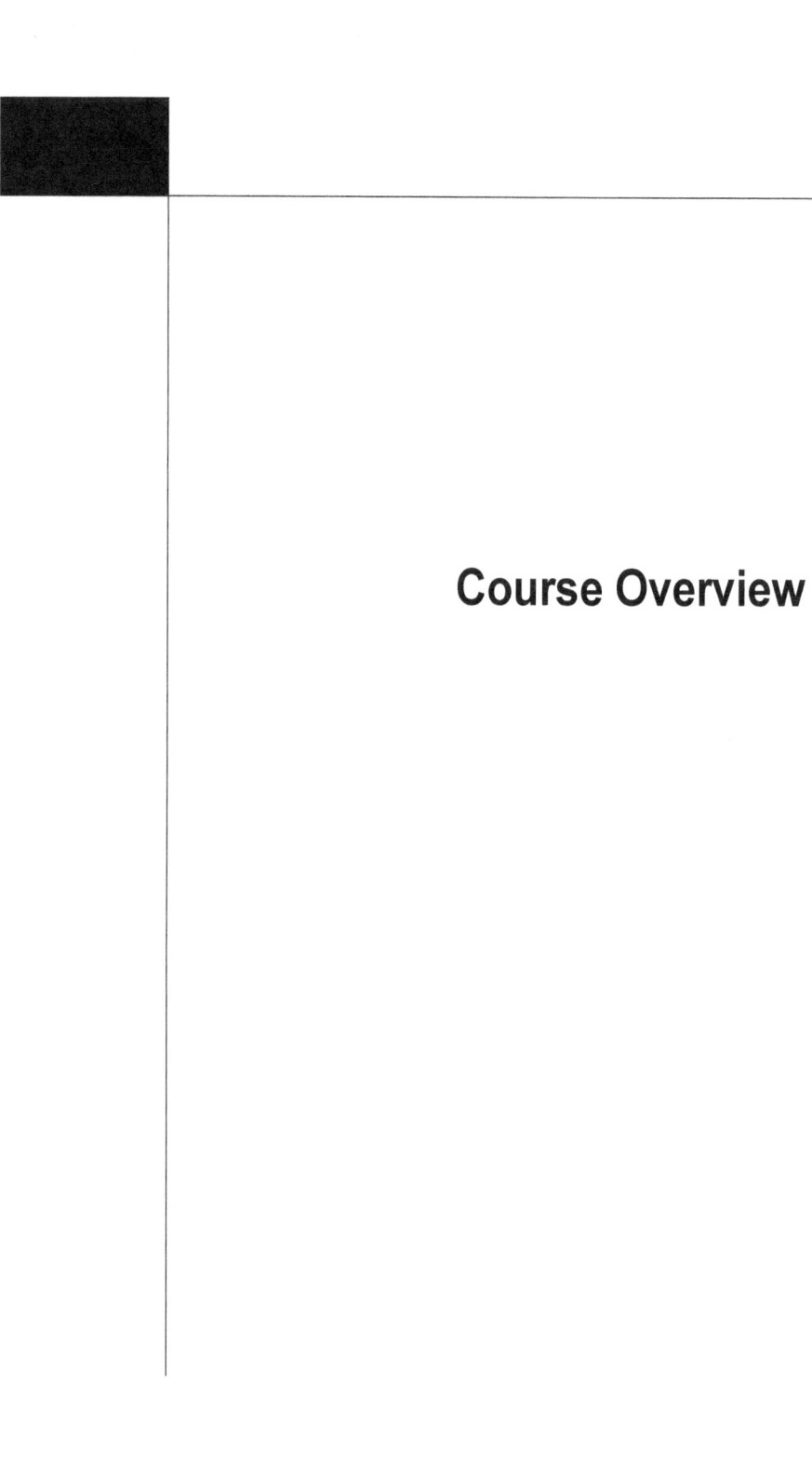

Course Overview

About This Course

This course is designed for emergency management personnel who are involved in developing an effective emergency planning system.

This course offers training in the fundamentals of the emergency planning process, including the rationale behind planning. It will develop your capability for effective participation in the all-hazard emergency operations planning process to save lives and protect property threatened by disaster.

FEMA's Independent Study Program

The Federal Emergency Management Agency's (FEMA's) Independent Study Program is one of the delivery channels that the Emergency Management Institute (EMI) uses to provide training to the general public and specific audiences. This course is part of FEMA's Independent Study Program. In addition to this course, the Independent Study Program includes other courses in the Professional Development Series (PDS) as well as courses in floodplain management, radiological emergency management, the role of the Emergency Manager, hazardous materials, disaster assistance, the role of the Emergency Operations Center (EOC), and an orientation to community disaster exercises.

FEMA's independent study courses are available at no charge and include a final examination. You may apply individually or through group enrollment. When enrolling for a course, you must include your name, mailing address, Social Security number, and the title of the course in which you wish to enroll.

If you need assistance with enrollment, or if you have questions about how to enroll, contact the Independent Study Program Administrative Office at:

FEMA Independent Study Program
Administrative Office
Emergency Management Institute
16825 South Seton Avenue
Emmitsburg, MD 21727
(301) 447-1200

FEMA's Independent Study Program (Continued)

Information about FEMA's Independent Study Program also is available on the Internet at:

http://www.training.fema.gov/IS

Each request will be reviewed and directed to the appropriate course manager or program office for assistance.

Course Completion

The course completion deadline for all FEMA Independent Study courses is 1 year from the date of enrollment. The date of enrollment is the date that the EMI Independent Study Office will use for completion of all required course work, including the final examination. If you do not complete this course, including the final examination, within that timeframe, your enrollment will be terminated.

Course Prerequisites

Emergency Planning has no prerequisites.

Final Examination

This course includes a final examination, which you must complete and return to FEMA's Independent Study Office for scoring. To obtain credit for taking this course, you must successfully complete this examination with a score of 75 percent or above. You may take the final examination as many times as necessary.

When you have completed all units, you must take the final examination online. EMI will score your test and notify you of the results.

Unit 1: Course Introduction

Introduction

The knowledge of how to plan for disasters is critical in emergency management. Planning can make a difference in mitigating against the effects of a disaster, including saving lives and protecting property, and helping a community recover more quickly from a disaster.

This course, *Emergency Planning*, is designed to aid emergency management personnel in developing an effective Emergency Operations Plan (EOP). Topics covered include selecting the planning team, the planning process, hazard analysis, and plan format. This course also prepares you to take the classroom course, *Workshop in Emergency Management* (WEM).

How To Take This Course

This independent study course is designed so that you can complete it on your own, at your own pace. Take a break after each unit, and give yourself time to think about the material, particularly as it applies to your work as an emergency management professional and the situations you have encountered or anticipate encountering on the job.

Emergency Planning contains seven units. Each of the units is described below.

- **Unit 1, Course Introduction,** provides an overview of the course objectives and instructions for taking the course.

- **Unit 2, The Planning Process,** provides an overview of the emergency planning process, including the steps involved, and how to determine who should be a part of the emergency planning team.

- **Unit 3, Threat Analysis,** describes the threat analysis process and explains why conducting a thorough threat analysis is a critical first step in emergency operations planning.

- **Unit 4, The Basic Plan,** introduces the purpose of the basic EOP and describes its components.

How To Take This Course (Continued)

- **Unit 5, Annexes and Appendices,** introduces functional annexes and hazard-, threat-, and incident-specific appendices, including their purposes and the differences between annexes and appendices.

- **Unit 6, Implementing Instructions,** introduces the different types of implementing instructions that may be developed at the agency level and how they are used.

- **Unit 7, Course Summary,** reviews and summarizes the course content and serves as preparation for the final exam.

Activities

This course will involve you actively as a learner by including activities that highlight basic concepts. Through the use of case studies, the course will also provide you with guidance on actions required in specific situations. These activities emphasize different learning points, so be sure to complete all of them. Compare your answers to the answers provided following the activity. If your answers are correct, continue on with the material. If any of your answers are incorrect, go back and review the material before continuing.

Knowledge Checks

To help you know when to proceed to the next unit, Units 2 through 6 are followed by a Knowledge Check that asks you to answer questions that pertain to the unit content, followed by the answers. When you finish each Knowledge Check, check your answers, and review the parts of the text that you do not understand. It would be to your benefit to be sure that you have mastered the current unit before proceeding to the next unit.

Appendixes

In addition to the course units, this course includes three appendixes. Appendix A contains a sample plan that will be used to complete an activity in Unit 4. Appendix B provides Job Aids, and Appendix C gives a list of the acronyms used in the course.

Final Examination

This course includes a written final examination, which you must complete and submit to FEMA's Independent Study Office. To obtain credit for taking this course, you must successfully complete this examination with a score of 75 percent or above. You may take the final examination as many times as necessary.

When you have completed all units, take the final examination online. EMI will score your test and notify you of the results.

Sample Learning Schedule

Complete this course at your own pace. You should be able to finish the entire course—including pretest, units, knowledge checks, and the final examination—in approximately 10 hours. The following learning schedule is an example intended to show relative times for each unit.

Unit	Suggested Time
Unit 1: Course Introduction	$^1/_2$ hour
Unit 2: The Planning Process	$1^1/_2$ hours
Unit 3: Threat Analysis	2 hours
Unit 4: The Basic Plan	2 hours
Unit 5: Annexes and Appendices	2 hours
Unit 6: Implementing Instructions	$^3/_4$ hour
Unit 7: Course Summary	$1^1/_4$ hours

Case Study: Why Plan?

Instructions: *Read the following case study. As you read, think about how the planning ability of these communities compares with that in your own community. Answer the questions that follow the case study. Then turn the page to check your answers against the answers provided.*

At 6:53 p.m., on Friday, October 6, Hurricane Frieda slammed into the Carolinas. A category 3 hurricane, Frieda dumped 12 inches of rain in as many hours, causing coastal flooding that, combined with wind speeds of 115 m.p.h., demolished 1,000 homes, seriously damaged 25,000 others, and left 150,000 people homeless. Mass evacuation in coastal counties was required.

Evacuation in most counties went well. Prior to the hurricane, Green County had conducted a study to estimate the time required to evacuate its population, and the actual time to evacuate was less than planned. Additionally, inland residents were able to survive on their own for several days, thanks to functioning county emergency services.

However, evacuation in Washington and Jefferson counties, which had no emergency plans, was itself a disaster. The decision to recommend evacuation was made too late and was broadcast insufficiently. Furthermore, evacuation routes were not specified. Traffic on westbound two-lane roads crawled to a standstill, and many drivers had to abandon their cars in rising water and proceed on foot in high winds. There were many casualties among those trying to reach shelter. These counties had to request State help immediately to rescue residents. After the storm, these counties were not eligible for the full amount of State aid to rebuild because of their failure to create an emergency plan.

1. What advantages to emergency planning can you list from this case study?

2. What consequences resulted from a lack of planning?

 **Case Study: Why Plan? (Continued)**

Answers to the Case Study

Advantages to counties with emergency plans were their ability to:

- *Evacuate successfully.*
- *Survive on their own for several days.*

Consequences to counties without emergency plans were:

- *Their need for immediate assistance.*
- *The casualties resulting from attempted evacuation.*
- *Their ineligibility for the full amount of State aid. (In most States, counties that do not have emergency plans cannot declare an emergency and are ineligible for any aid or for the full amount of aid.)*

The bottom line is that laws require counties to do everything reasonable and prudent to protect lives and property, including emergency planning.

In the space below, consider your own community, and list at least three benefits it could gain from having an up-to-date plan.

Course Goals

In *Emergency Planning*, you will learn how to plan for a disaster. This course will provide you with a foundation that will enable you to:

- Answer the question, "Why plan?"

- Describe the threat analysis and capability assessment process.

- Describe the EOP format and content.

- Identify types of community support available and required for response and recovery.

- State the rationale for a team approach to planning.

- Describe and demonstrate EOP coordination and marketing.

- Relate exercises to the planning process.

- Develop a plan maintenance program.

- Develop and present a personal action plan for emergency planning.

Goal Setting

What do you hope to gain through completing *Emergency Planning*?
Depending on your role in emergency management, your prior experience, and your current level of expertise, your goals may be slightly different from those of other emergency management professionals.

Clarifying your goals will help you gain the most from the time you spend completing this course. Take a few minutes to complete the following activity.

Activity: Personal Learning Goals

The purpose of this activity is to help you develop personal goals for this course. Consider the following information:

- *The course goals.*
- *Your own experience with emergency planning.*

Think about what you would like to accomplish through this course. Then list three (or more) personal goals for improving your ability to plan for an emergency.

Goals
1. _____
2. _____
3. _____

Unit Summary

Knowing how to plan for a disaster is critical because effective planning can make a difference in:

- Mitigating against a disaster's effects.

- Helping a community recover more quickly.

This course is designed to aid in developing an effective Emergency Operations Plan. You can complete this course on your own, at your own pace.

The course contains components that will guide you through the learning, including:

- **Activities** to provide guidance on actions required in specific situations.

- **Knowledge Checks** to test yourself on what you have learned and review the parts that you do not understand.

- **Appendix** that contains an acronym list.

Unit 1 gave you an overview of *Emergency Planning* and instructions on how to take the course. Unit 2 will examine the planning process.

For More Information

- Guide for All-Hazard Emergency Operations Planning (Comprehensive Preparedness Guide (CPG) 101):

  http://www.fema.gov/about/divisions/cpg.shtm

- Electronic Journal of Emergency Management (available free online):

  http://members.tripod.com/~Richmond_ESM/index.html

Unit 2: The Planning Process

Introduction and Unit Overview

This unit will provide an overview of the emergency planning process, including who should participate on the planning team. After you complete this unit, you should be able to:

- Describe the key steps in the emergency planning process.

- Identify agencies that should be involved in emergency planning.

- Describe where you fit into the emergency planning process.

Mandates: Incident Management and Coordination Systems

On February 28, 2003, the President issued Homeland Security Presidential Directive 5 (HSPD–5), "Management of Domestic Incidents," which directed the Secretary of Homeland Security to develop and administer a National Incident Management System (NIMS). This system provides a consistent nationwide template to enable Federal, State, tribal, and local governments, nongovernmental organizations (NGOs), and the private sector to work together to prevent, protect against, respond to, recover from, and mitigate the effects of incidents, regardless of cause, size, location, or complexity. This consistency provides the foundation for utilization of NIMS for all incidents, ranging from daily occurrences to incidents requiring a coordinated Federal response.

National Incident Management System (NIMS)

NIMS is not an operational incident management or resource allocation plan. NIMS represents a core set of doctrines, concepts, principles, terminology, and organizational processes that enables effective, efficient, and collaborative incident management.

Building on the foundation provided by existing emergency management and incident response systems used by jurisdictions, organizations, and functional disciplines at all levels, NIMS integrates best practices into a comprehensive framework for use nationwide by emergency management/response personnel in an all-hazards context. These best practices lay the groundwork for the components of NIMS and provide the mechanisms for the further development and refinement of supporting national standards, guidelines, protocols, systems, and technologies. NIMS fosters the development of specialized technologies that facilitate emergency management and incident response activities, and allows for the adoption of new approaches that will enable continuous refinement of the system over time.

NIMS (Continued)

Five major components make up the NIMS system approach:

- **Preparedness:** Effective emergency management and incident response activities begin with a host of preparedness activities conducted on an ongoing basis, in advance of any potential incident. Preparedness involves an integrated combination of assessment; planning; procedures and protocols; training and exercises; personnel qualifications, licensure, and certification; equipment certification; and evaluation and revision.

- **Communications and Information Management:** Emergency management and incident response activities rely on communications and information systems that provide a common operating picture to all command and coordination sites. NIMS describes the requirements necessary for a standardized framework for communications and emphasizes the need for a common operating picture. This component is based on the concepts of interoperability, reliability, scalability, and portability, as well as the resiliency and redundancy of communications and information systems.

- **Resource Management:** Resources (such as personnel, equipment, or supplies) are needed to support critical incident objectives. The flow of resources must be fluid and adaptable to the requirements of the incident. NIMS defines standardized mechanisms and establishes the resource management process to identify requirements, order and acquire, mobilize, track and report, recover and demobilize, reimburse, and inventory resources.

- **Command and Management:** The Command and Management component of NIMS is designed to enable effective and efficient incident management and coordination by providing a flexible, standardized incident management structure. The structure is based on three key organizational constructs: the Incident Command System, Multiagency Coordination Systems, and Public Information.

- **Ongoing Management and Maintenance:** Within the auspices of Ongoing Management and Maintenance, there are two components: the National Integration Center (NIC) and Supporting Technologies.

Additional information about NIMS can be accessed online at http://www.fema.gov/emergency/nims/ or by completing EMI's IS 700 online course.

National Response Framework (NRF)

The NRF is a guide to how the Nation conducts all-hazards response – from the smallest incident to the largest catastrophe. This key document establishes a comprehensive, national, all-hazards approach to domestic incident response. The Framework identifies the key response principles, roles, and structures that organize national response. It describes how communities, States, the Federal Government, and private-sector and nongovernmental partners apply these principles for a coordinated, effective national response.

The NRF is:

- **Always in effect, and elements can be implemented as needed on a flexible, scalable basis to improve response.** It is not always obvious at the outset whether a seemingly minor event might be the initial phase of a larger, rapidly growing threat. The NRF allows for the rapid acceleration of response efforts without the need for a formal trigger mechanism.

- **Part of a broader strategy.** The NRF is required by, and integrates under, a larger National Strategy for Homeland Security that:

  - Serves to guide, organize, and unify our Nation's homeland security efforts.

  - Reflects our increased understanding of the threats confronting the United States.

  - Incorporates lessons learned from exercises and real-world catastrophes.

  - Articulates how we should ensure our long-term success by strengthening the homeland security foundation we have built.

- **Comprised of more than the core document.** The NRF is comprised of the core document, the Emergency Support Function (ESF), Support, and Incident Annexes, and the Partner Guides. The core document describes the doctrine that guides our national response, roles and responsibilities, response actions, response organizations, and planning requirements to achieve an effective national response to any incident that occurs.

The following documents provide more detailed information to assist practitioners in implementing the Framework:

- **Emergency Support Function Annexes** group Federal resources and capabilities into functional areas that are most frequently needed in a national response (e.g., Transportation, Firefighting, Search and Rescue).

NRF (Continued)

- **Support Annexes** describe essential supporting aspects that are common to all incidents (e.g., Financial Management, Volunteer and Donations Management, Private-Sector Coordination).
- **Incident Annexes** address the unique aspects of how we respond to seven broad incident categories (e.g., Biological, Nuclear/Radiological, Cyber, Mass Evacuation).

Additional information about the NRF can be accessed online at http://www.fema.gov/emergency/NRF/ or by completing EMI's IS 800.b online course.

What This Means to You

Your jurisdiction is required to:

- Use NIMS to manage all incidents, including recurring and/or planned special events.

- Integrate all response agencies and entities into a single, seamless system, from the Incident Command Post, through department Emergency Operations Centers (DEOCs) and local Emergency Operations Centers (EOCs), through the State EOC to the regional- and national-level entities.

- Develop and implement a public information system.

- Identify and type all resources according to established standards.

- Ensure that all personnel are trained properly for the job(s) they perform.

- Ensure communications interoperability and redundancy.

Remember the importance of working with VOADs, NGOs, business and industry, and others to develop a plan for addressing volunteer needs *before* an incident to help eliminate some of the potential problems that can occur *during* an incident.

The Emergency Planning Process

Emergency planning is not a one-time event. Rather, it is a continual cycle of planning, training, exercising, and revision that takes place throughout the five phases of the emergency management cycle (preparedness, prevention, mitigation, response, and recovery).

The planning process does have one purpose—the development and maintenance of an up-to-date Emergency Operations Plan (EOP). An EOP can be defined as *a document maintained by various jurisdictional levels describing the plan for responding to a wide variety of potential hazards.*

Although the emergency planning process is cyclic, EOP development has a definite starting point.

There are six steps in the emergency planning process:

1. **Form a collaborative planning team.** Using a team or group approach helps organizations define their perception of the role they will play during an operation. One goal of using a planning team is to build and expand relationships that help bring creativity and innovation to planning during an event. This approach helps establish a planning routine, so that processes followed before an event occurs are the same as those used during an event.

2. **Understand the situation.** Hazards and threats are the general problems that jurisdictions face. Researching and analyzing information about potential hazards and threats a jurisdiction may face brings specificity to the planning process. If hazards and threats are viewed as problems and operational plans are the solution, then hazard and threat identification and analysis are key steps in the planning process.

3. **Determine goals and objectives.** By using information from the hazard profile developed as part of the analysis process, the planning team thinks about how the hazard or threat would evolve in the jurisdiction and what defines a successful operation. Starting with a given intensity for the hazard or threat, the team imagines an event's development from prevention and protection efforts, through initial warning (if available) to its impact on the jurisdiction (as identified through analysis) and its generation of specific consequences (e.g., collapsed buildings, loss of critical services or infrastructure, death, injury, or displacement).

4. **Develop the plan.** The same scenarios used during problem identification are used to develop potential courses of action. For example, some prevention and protection courses of action can be developed that may require a significant initial action (such as hardening a facility) or creation of an ongoing procedure (such as checking identity cards.). Planners consider the needs and demands, goals, and objectives to develop several response alternatives.

The Emergency Planning Process (Continued)

5. **Prepare, review, and approve the plan.** The planning team develops a rough draft of the base plan, functional or hazard annexes, or other parts of the plan as appropriate. As the planning team works through successive drafts, the members add necessary tables, charts, and other graphics. A final draft is prepared and circulated to organizations that have responsibilities for implementing the plan to obtain their comments.

6. **Refine and execute the plan.** Exercising the plan and evaluating its effectiveness involve using training and exercises and evaluation of actual events to determine whether the goals, objectives, decisions, actions, and timing outlined in the plan led to a successful response. Similarly, planners need to be aware of lessons and practices from other communities.

The planning process is all about stakeholders bringing their resources and strengths to the table to develop and reinforce a jurisdiction's emergency management and homeland security programs. Properly developed, supported, and executed operational plans are a direct result of an active and evolving program.

Who Should Be Involved?

Emergency planning is a team effort because disaster response requires coordination between many community agencies and organizations and different levels of government. Furthermore, different types of emergencies require different kinds of expertise and response capabilities. Thus, the first step in emergency planning is identification of all of the parties that should be involved.

Obviously, the specific individuals and organizations involved in response to an emergency will depend on the type of disaster. Law enforcement will probably have a role to play in most events, as will fire, Emergency Medical Services (EMS), voluntary agencies, and the media. On the other hand, hazardous materials (HazMat) personnel may or may not be involved in a given incident but should be involved in the planning process because they have specialized expertise that may be called on.

You will be determining the types of hazards that pose a risk to your community in Unit 3, Threat Analysis. In the meantime, think broadly. Most of the individuals and organizations listed on the next page have a role to play in planning for different types of emergencies.

Who Should Be Involved? (Continued)

Individuals/Organizations	What They Bring to the Planning Team
Senior Official (elected or appointed) or designee	• Support for the homeland security planning process. • Government intent by identifying planning goals and essential tasks. • Policy guidance and decision-making capability. • Authority to commit the jurisdiction's resources.
Emergency Manager or designee	• Knowledge about all-hazard planning techniques. • Knowledge about the interaction of the tactical, operational, and strategic response levels. • Knowledge about the prevention, protection, mitigation, response, and recovery strategies for the jurisdiction. • Knowledge about existing mitigation, emergency, continuity, and recovery plans.
EMS Director or designee	• Knowledge about emergency medical treatment requirements for a variety of situations • Knowledge about treatment facility capabilities • Specialized personnel and equipment resources • Knowledge about how EMS interacts with the Emergency Operations Center and incident command
Fire Chief or designee	• Knowledge about fire department procedures, on-scene safety requirements, hazardous materials response requirements, and search-and-rescue techniques. • Knowledge about the jurisdiction's fire-related risks. • Specialized personnel and equipment resources.
Police Chief or designee	• Knowledge about police department procedures; on-scene safety requirements; local laws and ordinances; explosive ordnance disposal methods; and specialized response requirements, such as perimeter control and evacuation procedures. • Knowledge about the prevention and protection strategies for the jurisdiction. • Knowledge about fusion centers and intelligence and security strategies for the jurisdiction. • Specialized personnel and equipment resources.
Public Works Director or designee	• Knowledge about the jurisdiction's road and utility infrastructure. • Specialized personnel and equipment resources.
Public Health Officer or designee	• Records of morbidity and mortality. • Knowledge about the jurisdiction's surge capacity. • Understanding of the special medical needs of the community. • Knowledge about historic infectious disease and syndromic surveillance.

	• Knowledge about infectious disease sampling procedures.

Who Should Be Involved? (Continued)

Individuals/Organizations	What They Bring to the Planning Team
Hazardous Materials Coordinator	• Knowledge about hazardous materials that are produced, stored, or transported in or through the community. • Knowledge about U.S. Environmental Protection Agency (EPA), Occupational Safety and Health Administration (OSHA), and U.S. Department of Transportation (DOT) requirements for producing, storing, and transporting hazardous materials and responding to hazardous materials incidents.
Hazard Mitigation Specialist	• Knowledge about all-hazard planning techniques. • Knowledge of current and proposed mitigation strategies. • Knowledge of available mitigation funding. • Knowledge of existing mitigation plans.
Transportation Director or designee	• Knowledge about the jurisdiction's road infrastructure. • Knowledge about the area's transportation resources. • Familiarity with the key local transportation providers. • Specialized personnel resources.
Agriculture Extension Service	• Knowledge about the area's agricultural sector and associated risks (e.g., fertilizer storage, hay and grain storage, fertilizer and/or excrement runoff).
School Superintendent or designee	• Knowledge about school facilities. • Knowledge about the hazards that directly affect schools. • Specialized personnel and equipment resources (e.g., buses).
Social services agency representatives	• Knowledge about special-needs populations
Local Federal asset representatives	• Knowledge about specialized personnel and equipment resources that could be used in an emergency. • Facility security and response plans (to be integrated with the jurisdiction's EOP). • Knowledge about potential threats to or hazards at Federal facilities (e.g., research laboratories, military installations).
NGOs (includes members of National VOAD [Voluntary Organizations Active in Disaster]) and other private, not-for-profit, faith-based, and community organizations	• Knowledge about specialized resources that can be brought to bear in an emergency. • Lists of shelters, feeding centers, and distribution centers. • Knowledge about special-needs populations.
Local business and industry representatives	• Knowledge about hazardous materials that are produced, stored, and/or transported in or through the community. • Facility response plans (to be integrated with the jurisdiction's EOP). • Knowledge about specialized facilities, personnel, and equipment resources that could be used in an emergency.
Amateur Radio Emergency Service	• List of ARES/RACES resources that can be used in an

(ARES)/Radio Amateur Civil Emergency Services (RACES) Coordinator	emergency.

Who Should Be Involved? (Continued)

Individuals/Organizations	What They Bring to the Planning Team
Utility representatives	• Knowledge about utility infrastructures. • Knowledge about specialized personnel and equipment resources that could be used in an emergency.
Veterinarians/animal shelter representatives	• Knowledge about the special response needs for animals, including livestock.

Expanded planning teams should include representatives from partners within the identified planning area, surrounding jurisdictions, and facilities or locations of concern and must include stakeholder organizations responsible for infrastructure, the economy, the environment, and quality of life.

Such organizations include those responsible for:

• Building codes.
• Land use and zoning.
• Transportation corridors.
• Utilities.
• Economic development.

How To Get the Team Together

Getting everyone to take an active interest in emergency planning will be no easy task. To schedule meetings with so many participants may be even more difficult. It is critical, however, to have everyone's participation in the planning process and to have them feel ownership in the plan by involving them from the beginning. Also, their expertise and knowledge of their organizations' resources is crucial to developing an accurate plan that considers the entire community's needs and the resources that could be made available in an emergency.

It is definitely to the community's benefit to have the active participation of all players. But what can you do to ensure that everyone who should participate does? Some tips for gathering the team together include:

- Planning ahead. Give the planning team plenty of notice of where and when the planning meeting will be held. If time permits, you might even survey the team members to find the time and place that will work for them.

- Providing information about team expectations. Explain why participating on the planning team is important to the participants' agencies and to the community itself. Show the participants how they will contribute to a more effective emergency response.

- Asking the senior elected or appointed official to sign the meeting announcement. A directive from the executive office will carry the authority of the CEO and send a clear signal that the participants are expected to attend and that emergency planning is important to the community.

- Allowing flexibility in scheduling after the first meeting. Not all team members will need to attend all meetings. Some of the work can be completed by task forces or subcommittees. Where this is the case, gain concurrence on timeframes and milestones but let the subcommittee members determine when it is most convenient to meet.

- Considering using external facilitators. Third-party facilitators can perform a vital function by keeping the process focused and mediating disagreements.

Also, talk to some Emergency Managers from adjacent communities to gain their ideas and input on how to gain and maintain interest in the planning process.

How Should the Team Operate?

Working with personnel from other agencies and organizations requires collaboration. **Collaboration** is the *process by which people work together as a team toward a common goal*—in this case, development of a community EOP.

Successful collaboration requires:

- A commitment to participate in *shared decisionmaking.*

- A *willingness to share* information, resources, and tasks.

- A professional sense of *respect* for individual team members.

Collaboration can be made difficult by differences among agencies and organizations in:

- Terminology.

- Experience.

- Mission.

- Culture.

Collaboration under these conditions requires the *flexibility* to reach agreement on common terms and priorities, and *humility* to learn from others' ways of doing things.

Collaboration among EOP planning team members benefits the community by strengthening the overall response to the disaster. For example, collaboration can:

- Eliminate duplication of services, resulting in a more efficient response.

- Expand resource availability.

- Enhance problem solving through cross-pollination of ideas.

How Should the Team Operate? (Continued)

However, collaboration does not come automatically. Building a team that works well together takes time and effort and typically evolves through the following stages:

1. Forming: Individuals come together as a team. During this stage the team members may be unfamiliar with each other and uncertain of their roles on the team.

2. Storming: Team members become impatient, disillusioned, and may disagree.

3. Norming: Team members accept their roles and focus on the process.

4. Performing: Team members work well together and make progress toward the goal.

5. Adjourning: Their task accomplished, team members may feel pride in their achievement and some sadness that the experience is ending. (Note that the planning team should never really adjourn. They will meet less frequently as they plan and conduct exercises and revise the plan, but the core of the team should remain intact.)

Team Roles

The team leader plays a key role in the development of effective teams through all stages of the team's development. The team leader can initiate appropriate team-building activities that move the team through the stages and toward its goal.

Other team roles besides the team leader include:

- Task Master: Identifies the work to be done and motivates the team.

- Innovator: Generates original ways to get the group's work done.

- Organizer: Helps groups develop plans for getting the work done.

- Evaluator: Analyzes ideas, suggestions, and plans made by the group.

- Finisher: Follows through on plans developed by the team.

Team Roles (Continued)

You will know your EOP planning team is on track when it displays the following characteristics:

- A common goal (i.e., development of an EOP)

- A leader who provides direction and guidance

- Open communication

- Constructive conflict resolution

- Mutual trust

- Respect for each individual and his or her contributions

The activity that follows asks you to think about what your contribution to the EOP planning team might be.

Activity: Organizational Roles and Individual Skills

This activity will give you the opportunity to think about what role you play in the emergency planning process. Answer the questions below.

1. What is your organization's role in emergency response?

2. What are your current emergency management responsibilities?

3. What other expertise do you have that could be useful to the emergency planning process (e.g., group facilitation skills, knowledge of building methods and materials, ability to simplify complex information so that it is readily understandable by a layperson)? Think expansively.

4. What do your skills, knowledge, and abilities contribute to the process?

Unit Summary

Emergency planning is a cycle of planning, training, exercising, and revision that continues throughout the five phases of the emergency management cycle (preparedness, prevention, response, recovery, and mitigation).

One purpose of the planning process is the development and maintenance of an up-to-date EOP. The six steps of the planning process are:

1. Form a collaborative planning team.

2. Understand the situation.

3. Determine goals and objectives.

4. Develop the plan.

5. Prepare, review, and approve the plan.

6. Refine and execute the plan.

Emergency planning is a team effort and requires collaboration with personnel from other agencies and organizations. Building an effective team takes time and effort as members go through several stages. The team leader plays a key role.

An effective EOP planning team displays the following characteristics:

- A common goal (development of the EOP)

- A leader who provides direction and guidance

- Open communication

- Constructive conflict resolution

- Mutual trust

- Respect for each individual and his or her contributions

Think about your role in the planning process—how you fit in.

This unit provided an overview of the emergency planning process and described who should participate on the planning team, how to get everyone involved, and how to get everyone working toward a common goal. Unit 3 will discuss the hazard analysis process.

Knowledge Check

Carefully read each question and all of the possible answers before selecting the most appropriate response for each test item. Fill in the blank, or circle the letter corresponding to the answer you have chosen.

1. The first step in the emergency planning process is _____.

2. The second stage in team-building is performing.

   a. True
   b. False

3. Which of the following team roles is key in motivating the team?

   a. Leader
   b. Organizer
   c. Evaluator
   d. Finisher

4. Which of the following is <u>not</u> a characteristic of effective teams?

   a. Open communication
   b. Constructive conflict resolution
   c. Mutual trust
   d. A preference for certain members' contributions over others

5. Collaboration among agencies is _____ by differing experiences.

   a. Enhanced
   b. Made more difficult

Knowledge Check (Continued)

1. Form the planning team.
2. b
3. a
4. d
5. b

Unit 3: Threat Analysis

Introduction and Unit Overview

In Unit 2, you learned about the planning process and about forming a planning team. In this lesson, you'll learn about the second step of the planning process – understanding the situation.

In this unit, you will be introduced to research and analysis tools that can be used to:

- Identify potential threats to your community.

- Develop hazard profiles.

- Quantify and prioritize risks.

The Threat Analysis Process

Threat analysis determines:

- What can occur.

- How often it is likely to occur.

- The damage it is likely to cause.

- How it is likely to affect the community.

- How vulnerable the community is to the threat.

The Threat Analysis Process (Continued)

The steps in the threat analysis process are:

1. Identify threats.

2. Profile each threat.

3. Develop a community profile.

4. Determine vulnerability.

5. Create and apply scenarios.

Each of these steps will be covered in the sections that follow.

Step 1: Identifying Threat

The first step is to develop a list of threats the community may face. This list usually is based on historical data about past events. Information about recent events is relatively easy to gather, while information about older events may be more difficult to find. Threats can be:

- **Natural.** Natural threats tend to occur repeatedly in the same geographical locations because they are related to weather patterns and/or physical characteristics of an area. Examples include: severe weather, fire, drought, typhoons, epidemics, etc.

- **Technological.** Technological threats originate from technological or industrial accidents, infrastructure failures, or certain human activities. Technological threats may include: cyber/database failures, urban fires, radiological or hazardous material releases, power failures, transportation accidents, dam failures, bridge collapses, etc.

- **Human-caused.** Human-caused threats arise from deliberate, intentional human actions to threaten or harm the well-being of others. Human-caused threats may include: kidnappings, hostage situations, sabotage, civil disturbances, bombings, hijacking, terrorist acts, etc.

Step 1: Identifying Threats (Continued)

There are many potential sources of threats information. Some sources, such as local newspapers, are fairly obvious. To get a more complete picture of the types of hazards that a community has faced historically, it may be necessary to check other sources, such as:

- The State Department of Agriculture, Bureau of Labor Statistics, or other agencies.

- The National Weather Service (NWS).

- Local historical societies.

- Anecdotal information from long-time residents.

There may be other sources of information, and you should take some time to check them so that your threat analysis is as complete as possible.

If your community has an existing threat analysis, the best way to begin is by reviewing it and identifying any changes that may have occurred since it was developed or updated last.

Some possible changes within or near the community that could cause threat analysis information to change over time include:

- New mitigation measures (e.g., stronger building codes, addition of roof or foundation braces).

- The opening or closing of facilities or structures that pose potential secondary hazards (e.g., hazardous materials facilities and transport routes).

- Local development activities.

- Climatic changes.

- Terrorist threats.

There may be other long-term changes to investigate as well. These changes, such as climatic changes in average temperature or rainfall/snowfall amounts, are harder to track but could be very important to the hazard analysis.

Step 1: Identifying Threats (Continued)

The following is a sample threat list:

Natural Threats	Technological Threats	Human-Caused Threats
• Avalanche • Drought • Earthquake • Epidemic • Flood • Hurricane • Landslide • Tornado • Tsunami • Volcanic eruption • Wildfire • Winter storm	• Airplane crash • Dam failure • HAZMAT release • Power failure • Radiological release • Train derailment • Urban conflagration	• Civil disturbance • School violence • Terrorist act • Sabotage

Step 2: Profiling Threats

Threat profiles should address each threat's:

- Type.

- Probability of occurring.

- Past history.

- Potential consequences.

The availability of warnings will also become a crucial part of the threat profile.

A threat-analysis worksheet will help you identify and prioritize threats. Below is an example of completed sections from the worksheet:

Threat: *Seasonal Floods*

Potential Consequences:
☐ **Catastrophic** (Mass fatalities/casualties, loss of governance and essential services, widespread damages)
☐ **Severe** (Numerous fatalities/ casualties, loss of essential services, and widespread damage)
☐ **Moderate** (Limited number of fatalities/casualties and damage to property)
☑ **Minor** (Little or no injuries and isolated damage)

Probability of Occurring:	Past History:
☑ High ☐ Medium ☐ Low	Has this type of incident occurred before? ☑ Yes ☐ No If yes, when? *6 months ago*

Activity: Profiling a Threat

This activity will give you an opportunity to complete a simple analysis of a threat that could affect your community. To complete this activity, think of a threat that your community has faced in the past, or is predicted to occur in the future, and write that threat on the line below. (Note: Do not limit yourself to natural hazards.)

Threat: _____

1. Think of several sources of information that you could check to investigate this threat.

2. Complete the threat profiling activity on the following pages to develop a first draft of a threat profile for the threat you chose. (Note: To complete the profile, you will need to consult your sources and gather all needed information on this threat.)

Activity: Profiling a Threat (Continued)

Threat Profile Worksheet
Threat:
Potential Consequences: ☐ Catastrophic (Mass fatalities/casualties, loss of governance and essential services, widespread damages) ☐ Severe (Numerous fatalities/ casualties, loss of essential services, and widespread damage) ☐ Moderate (Limited number of fatalities/casualties and damage to property) ☐ Minor (Little or no injuries and isolated damage)

Probability of Occurring: ☐ High ☐ Medium ☐ Low	Past History: Has this type of incident occurred before? ☐ Yes ☐ No If yes, when? _____

Areas Likely to be Affected Most:
Probable Duration:
Potential Speed of Onset (Probable amount of warning time): • Minimal (or no) warning.   • 12 to 24 hours warning. • 6 to 12 hours warning.   • More than 24 hours warning.
Existing Population Warning Systems:
Does a Vulnerability Analysis Exist? Yes ☐ No ☐

Note that some threats may pose such a limited risk to the community that additional analysis is not necessary.

Step 3: Developing a Community Profile

The next step in the threat analysis process is to combine threat-specific information with a profile of your community to determine the community's vulnerability to the threat (or risk of damage from the threat).

Because different communities have different profiles, vulnerabilities to the same disaster will vary.

The table below summarizes key factors that are included in a community profile.

Key Community Factors				
Geography	Property	Infrastructure	Demographics	Response Organizations
• Major geographic features • Typical weather patterns	• Numbers • Types • Ages • Building codes • Critical facilities • Potential secondary hazards	• Utilities construction, layout, access • Communication system layout, features, backups • Road systems • Air and water support	• Population size, distribution, concentrations • Numbers of people in vulnerable zones • Special populations • Animal populations	• Locations • Points of contact • Facilities • Services • Resources

After gathering this information about the community, emergency managers use it to develop the community's threat analysis, as shown in the table on the next page.

Step 3: Developing a Community Profile (Continued)

Use of Community Factors in Threat Analysis	
Type of Information	Used In:
Geographic	• Predicting risk factors and the impact of potential hazards and secondary hazards.
Property	• Projecting consequences of potential hazards to the local area. • Identifying available resources (e.g., for sheltering).
Infrastructure	• Identifying points of vulnerability. • Preparing evacuation routes, emergency communications, and project response and recovery requirements.
Demographic	• Projecting consequences of disasters on the population. • Disseminating warnings and public information. • Planning evacuation and mass care.
Response Organizations	• Identifying response capabilities.

Step 4: Determining Vulnerability

After community and threat profiles have been compiled, it is helpful to quantify the community's risk by merging the information so that the community can focus on the hazards that present the highest risk.

Risk is the *predicted impact that a hazard would have on the people, services, and specific facilities in the community.* For example, during heavy rains, a specific road might be at risk of flooding, leading to restricted access to a critical facility.

Quantifying risk involves:

- Identifying the elements of the community (populations, facilities, and equipment) that are potentially at risk from a specific threat.

- Developing response priorities.

- Assigning severity ratings.

- Compiling risk data into community risk profiles.

In surveying risk, it is helpful to develop response priorities. The following is a suggested hierarchy for setting priorities:

- Priority 1: Life safety (including hazard areas, high-risk populations, and potential search and rescue situations). Keep in mind that response personnel cannot respond if their own facilities are affected.

- Priority 2: Essential facilities.

- Priority 3: Critical infrastructure (utilities, communication, and transportation systems).

Prioritizing Risks

The community should assign each hazard a **severity rating**—or **risk index**—that will *predict, to the degree possible, the damage that can be expected in that community as a result of that threat.*

This rating quantifies the expected impact of a specific hazard on people, essential facilities, property, and response assets.

The following is an example of severity ratings that may be used:

Severity	Characteristics
Catastrophic	<ul><li>Multiple deaths.</li><li>Complete shutdown of critical facilities for 30 days or more.</li><li>More than 50 percent of property severely damaged.</li></ul>
Critical	<ul><li>Injuries and/or illnesses result in permanent disability.</li><li>Complete shutdown of critical facilities for at least 2 weeks.</li><li>More than 25 percent of property is severely damaged.</li></ul>
Limited	<ul><li>Injuries and/or illnesses do not result in permanent disability.</li><li>Complete shutdown of critical facilities for more than 1 week.</li><li>More than 10 percent of property is severely damaged.</li></ul>
Negligible	<ul><li>Injuries and/or illness treatable with first aid.</li><li>Minor quality of life lost.</li><li>Shutdown of critical facilities and services for 24 hours or less.</li><li>Less than 10 percent of property severely damaged.</li></ul>

Prioritizing Risks (Continued)

Next, develop a risk index for each threat by assigning a value to each severity level (use the following values: 1 = catastrophic; 2 = critical; 3 = limited; 4 = negligible) for the following types of threat data.

- Magnitude.

- Frequency of occurrence.

- Speed of onset.

- Community impact (severity rating).

- Special characteristics.

Finally, average the severity level for each factor to determine the overall risk level for that threat.

Activity: Prioritizing Risks

This activity will provide you with an opportunity to develop a risk index for the threat you identified earlier in this unit. To complete this activity, assign a severity level to each type of threat data. Then, average the severity level to determine the overall risk to your community.

Hazard: _____

Characteristic	Severity
Magnitude	1. Catastrophic 2. Critical 3. Limited 4. Negligible
Frequency of Occurrence	1. Catastrophic 2. Critical 3. Limited 4. Negligible
Speed of Onset	1. Catastrophic 2. Critical 3. Limited 4. Negligible
Community Impact	1. Catastrophic 2. Critical 3. Limited 4. Negligible
Special Characteristics	1. Catastrophic 2. Critical 3. Limited 4. Negligible
Total Risk	1. Catastrophic 2. Critical 3. Limited 4. Negligible

Note: When you prioritize all of the risks for your community, it will be easier if you develop a form that shows all of the threats on one or two pages.

Step 5: Creating and Applying Scenarios

The final step in the threat analysis process is to develop scenarios for the top-ranked threats (or those that rank above a specified threshold) that lay out the threat's development into an emergency. Scenarios should be realistic and based on the community's threat and risk data.

To create a scenario, emergency managers brainstorm to track the development of a specific type of emergency. A scenario should describe:

- The initial warning of the event.

- The potential overall impact on the community.

- The potential impact of the event on specific community sectors.

- The potential consequences, such as damage, casualties, and loss of services.

- The actions and resources that would be needed to deal with the situation.

Creating scenarios helps to identify situations that may exist in a disaster. These situations should be used to help ensure that your community is prepared should the threat event occur.

Activity: Threat Analysis

Follow the steps below to complete this activity.

1. Imagine a scenario based on the threat you profiled earlier. (Note: The scenario may not give all of the information needed to completely fill out some of the charts.)

2. Answer the questions on the following pages.

Activity: Threat Analysis (Continued)

3. Using the table below as a guide, make notes in the space below about the key community factors in your community.

Geography	Property	Infrastructure	Demographics	Response Organizations
Major geographic features:	Numbers:	Utilities construction, layout, access:	Population size, distribution, concentrations:	Locations:
	Types:	Communication system layout, features, backups:	Numbers of people in vulnerable zones:	Points of contact:
	Ages:			
Typical weather patterns:	Building codes:	Road systems:	Special populations:	Facilities:
	Critical facilities:			Services:
	Potential secondary hazards:	Air and water support:	Animal populations:	Resources:

Activity: Threat Analysis (Continued)

4. Develop a risk index for the hazard by assigning a value to each characteristic (using the following values: 1 = catastrophic; 2 = critical; 3 = limited; 4 = negligible) for the following types of threat data: (**Note:** Ignore frequency of occurrence for this exercise.)

- Magnitude

- Frequency of occurrence

- Speed of onset

- Community impact (severity rating)

- Special characteristics

Finally, average the values to arrive at a single risk index figure for the threat event.

Unit Summary

Threat analysis determines:

- What can occur.

- How often it is likely to occur.

- The damage it is likely to cause.

- How it is likely to affect the community.

- How vulnerable the community is to the threat.

There are 5 steps in the threat analysis process:

1. Identify threats.

2. Profile each threat.

3. Develop a community profile.

4. Determine vulnerability.

5. Create and apply scenarios.

In this unit, you have learned ways to identify what threats can occur in your community, the likelihood and severity of specific threats, and community vulnerability to various threats. Unit 4 will discuss the components of a basic EOP.

For More Information

- HAZUS: National Hazard Loss Estimating Methodology:

  http://www.fema.gov/plan/prevent/hazus/index.shtm

- Understanding Your Risks: Identifying Hazards and Estimating Losses (FEMA 386-2). Step-by-step guidance on how to accomplish a risk assessment:

  http://www.fema.gov/library/viewRecord.do?id=1880

- The Natural Hazards Center, University of Colorado, Boulder. Continuously updated comprehensive list of links related to natural hazards:

  www.colorado.edu/hazards

 **Knowledge Check**

Carefully read each question and all of the possible answers before selecting the most appropriate response for each test item. Circle the letter corresponding to the answer you have chosen.

1. The first step in a threat analysis is to:

   a. Divide the community into emergency management sectors.
   b. Create scenarios to test response capabilities.
   c. Develop a list of threats your community may face.
   d. Quantify the community's risks from identified threats.

2. A community profile should include information about the community's:

   a. Voting patterns.
   b. Congressional delegation(s).
   c. Local ordinances.
   d. Geography.

3. Information is collected about the community infrastructure in order to identify response capabilities.

   a. True
   b. False

4. Threat profiles should address each threat's:

   a. Sector population.
   b. Quantification of risk.
   c. Seasonal pattern.
   d. Event severity.

5. Risk is the predicted impact that a threat would have on people, services, and specific facilities and structures in the community.

   a. True
   b. False

Knowledge Check (Continued)

1. c
2. d
3. b
4. c
5. a

Unit 4: The Basic Plan

Introduction and Unit Overview

This unit will introduce and describe the purpose of the basic plan. After completing this unit, you should be able to:

- Identify the parts of an EOP.

- List the components of a basic plan.

- Describe the purpose of each component of the basic plan.

Components of a Basic Plan

Let's begin by reviewing our definition of an EOP from Unit 2. An **EOP** is a *document describing how citizens, property, and the environment will be protected in a disaster or emergency.*

The EOP describes actions to be taken in response to natural, manmade, or technological hazards, detailing the tasks to be performed by specific organizational elements at projected times and places based on established objectives, assumptions, and assessment of capabilities.

An EOP should be:

- Comprehensive. It should cover all aspects of emergency prevention, preparedness, and response and address mitigation concerns as well.

- All-hazard in approach and, thus, *flexible* enough to use in all emergencies—even unforeseen events.

- Risk-based. It should include hazard-specific information, based on the hazard analysis.

Components of a Basic Plan (Continued)

The purposes of the basic plan are to:

- Give an overview of the community's emergency response organization and policies.

- Provide a general understanding of the community's approach to emergency response for all involved agencies and organizations.

Although the basic plan provides the general approach to emergency response, it does not stand by itself. Rather, it forms the basis for the remainder of the plan, which also includes:

- Functional annexes that address the performance of a particular broad task or function, such as mass care or communications.

- Hazard-specific appendices that provide additional information specific to a particular hazard.

Annexes and appendices will be discussed in the next unit.

In addition, each part of the EOP may have addenda in the form of Standard Operating Procedures (SOPs), maps, charts, tables, forms, checklists, etc. These addenda may be included as attachments or incorporated by reference.

Components of a Basic Plan (Continued)

Although there is no mandatory format, the recommended format (for the sake of compatibility with other jurisdictions and levels of government) for the local basic plan includes the following components:

- Introduction. The introductory material consists of the following elements:

    - Promulgation document. The Promulgation document is signed by the jurisdiction's CEO, affirming his or her support for the emergency management agency and planning process. It gives organizations the authority and responsibility to perform their tasks. It also mentions the tasked organizations' responsibility to prepare and maintain implementing instructions, gives notice of necessary EOP revisions, and commits to the training necessary to support the EOP.
    - Signature page, which is signed by all partner organizations, demonstrating their commitment to EOP implementation.
    - Dated title page and record of changes (date, description, and affected parts) to the EOP.
    - Record of distribution, which lists EOP recipients, facilitating and giving evidence of EOP distribution. (EOP copies should be numbered and recorded.)
    - Table of contents.

- Purpose Statement, which includes:

    - A broad statement of what the EOP is meant to do.
    - A synopsis of the EOP, annexes, and appendices.

    The Purpose Statement need not be complex but should include enough information to establish the direction for the remainder of the plan.

- Scope. The operations plan should also explicitly state the scope of emergency and disaster response to which the plan applies and the entities (departments, agencies, private sector, citizens, etc.) and geographic areas to which it applies.

Components of a Basic Plan (Continued)

- Situation Overview. The Situation characterizes the planning environment, making clear why emergency operations planning is necessary. It draws from the threat analysis to narrow the scope of the EOP and includes the following types of information:

  - Hazards addressed by the plan
  - Relative probability and impact
  - Areas likely to be affected
  - Vulnerable critical facilities
  - Population distribution
  - Special-needs populations
  - Interjurisdictional relationships
  - Maps

- Planning Assumptions. The Assumptions statement delineates what was assumed to be true when the EOP was developed. The Assumptions statement shows the limits of the EOP, limiting liability. It may be helpful to list even "obvious" assumptions, such as:

  - Identified hazards will occur.
  - Individuals and organizations are familiar with the EOP.
  - Individuals and organizations will execute their assigned responsibilities.
  - Assistance may be needed and, if so, will be available.
  - Executing the EOP will save lives and reduce damage.

- Concept of Operations. This section explains the community's overall approach to emergency response (i.e., what, when, by whom). It includes:

  - The division of local, State, and Federal responsibilities.
  - When the EOP will be activated—and when it will be inactivated.
  - Alert levels and the basic actions that accompany each level.
  - The general sequence of actions before, during, and after an event.
  - Forms necessary to request assistance of various types.

- Organization and Assignment of Responsibilities. This section:

  - Lists the general areas of responsibility assigned by organization and position.
  - Identifies shared responsibilities (and specifies which organization has primary responsibility and which have supportive roles).

  The Organization and Assignment of Responsibilities section specifies reporting relationships and lines of authority for an emergency response. In addition, this section is where a jurisdiction discusses the response organizing option that it uses for emergency management – ESF, or agency and department, or functional areas of ICS/NIMS, or a hybrid.

Components of a Basic Plan (Continued)

- Direction, Control, and Coordination. This section:

    - Describes the framework for all direction, control, and coordination activities.
    - Identifies who has tactical and operational control of response assets.
    - Discusses multiagency coordination systems and processes used during an emergency.
    - Provides information on how department and agency plans nest into the EOP (horizontal coordination) and how higher-level plans are expected to layer on the EOP (vertical integration).

- Information Collection and Dissemination. This section:

    - Describes the required critical or essential information common to all operations identified during the planning process.
    - Identifies the type of information needed, where it is expected to come from, who uses the information, how the information is shared, the format for providing the information, and any specific times the information is needed.
    - May be expanded as an annex, or it may be included as an appendix or tab in the Direction, Control, and Coordination annex.

- Communications. This section:

    - Describes the response organization-to-response organization communication protocols and coordination procedures used during emergencies and disasters.
    - Discusses the framework for delivering communications support and how the jurisdiction's communications integrate into the regional or national disaster communications network.
    - May be expanded as an annex and is usually supplemented by communications SOPs and field guides.

Components of a Basic Plan (Continued)

- Administration, Finance, and Logistics. This section includes:

  - Assumed resource needs for high-risk hazards.
  - Resource availability.
  - Mutual aid agreements and assistance agreements.
  - Policies on augmenting response staff with public employees and volunteers.
  - A statement that addresses liability issues.
  - Resource management policies (acquisition, tracking, and financial recordkeeping).

- Plan Development and Maintenance. Responsibility for the coordination of the development and revision of the basic plan, annexes, appendices, and implementing instructions must be assigned to the appropriate persons. This section, therefore:

  - Describes the planning process.
  - Identifies the planning participants.
  - Assigns planning responsibilities.
  - Describes the revision cycle (i.e., training, exercising, review of lessons learned, and revision).

- Authorities and References. This section cites:

  - The legal bases for emergency operations and activities, including: laws, statutes, ordinances, executive orders, regulations, formal agreements, and predelegation of emergency authorities.
  - Pertinent reference materials (including related plans of other levels of government).

Activity: Basic Plan Review

This activity will give you the opportunity to review an EOP. You will find Section 1: Basic Plan from Jefferson County, Alabama's Comprehensive Emergency Management Plan, in Appendix A. Refer to the Jefferson County plan to answer the questions below. The answers follow the activity.

Organization and Assignment of Responsibilities

1. The plan specifies use of the Incident Command organization and structure at incident sites.

   a. Yes
   b. No

2. In the Specific Local Government Responsibilities, the Emergency Management Council responsibilities include: (Check all that apply.)

   a. Extending or terminating emergency/disaster declarations.
   b. Maintaining 800MHz radio system.
   c. Ensuring payroll system setup to pay employees.
   d. Providing overall direction and control.
   e. Making disaster declarations and requesting State and Federal assistance.

Concept of Operations

3. For a particular disaster event, the _____ decides which functional annexes are activated to meet the disaster response needs.

Administration and Logistics

4. The plan specifies that administrative procedures may not be relaxed, suspended, or made optional to respond to emergencies.

   a. Yes
   b. No

Activity: Basic Plan Review (Continued)

Direction and Control

5. Write the letter of the emergency level in the left-hand column next to the matching description in the right-hand column.

A. Level One Emergency

B. Level Two Emergency

C. Level Three Emergency

D. Level Four Emergency

_____ Non-routine assistance is required or anticipated, and the EOC is activated.

_____ Incidents at this level are common, and the responding department is responsible for controlling them.

_____ State and Federal assistance will be requested and is required.

_____ Internal and/or external agencies provide routine assistance, including mutual aid. Command and control remain the responsibility of the primary response department.

6. In G. Facilities, Emergency Operations Center, jurisdictions are encouraged to establish _____ to link to the county EOC via radio or telephone.

Attachment A: Primary Support Matrix

7. After heavy rains, a sewage treatment plant was inundated and pumping stations quit operating. The community water supply is contaminated. Annexes that definitely would apply to this situation include: (Circle all that apply.)

a. Annex 2, Situation Reporting
b. Annex 11, Health Services
c. Annex 15, Law Enforcement Services
d. Annex 13, Fire Services
e. Annex 10, Public Works

Activity: Basic Plan Review (Continued)

Suggested Answers:

1. a
2. a, d, and e
3. EMA Coordinator
4. b
5. c, a, d, b
6. On-scene command post
7. a, b, e

Activity: Reviewing Your Community's Basic Plan

Now that you have reviewed the contents and organization of a sample basic plan, you should review your community's plan to see what it contains and how it is organized. To complete this activity, obtain a copy of your community's basic plan and review it to see what types of information it contains and how it is organized. The questions below will help you organize your review.

Be aware that your community's plan may not be organized as described in this unit. Some States require a different structure. The important point to remember in your review is that, to be effective, the plan must include the types of content described in this unit.

1. Does your community's plan include a promulgation document signed by the CEO?

   ☐ Yes
   ☐ No

2. What organizations are signatories to the plan?

3. What threats or hazards are addressed by the plan?

4. Who has the authority to activate part or all of the plan?

Activity: Reviewing Your Community's Basic Plan (Continued)

5. With what jurisdictions does your community have mutual aid agreements or assistance agreements?

6. When was the last time that the plan was revised?

Unit Summary

The EOP describes actions to be taken in response to hazards. It details the tasks to be performed, specifies which organizational elements perform the tasks, and at what times and places.

An EOP should be:

- Comprehensive.

- All-hazard.

- Risk-informed.

The components of the basic plan include:

- Introductory Materials.

- Purpose Statement, Scope, Situation, and Assumptions.

- Concept of Operations.

- Organization and Assignment of Responsibilities.

- Direction, Control, and Coordination.

- Information Collection and Dissemination.

- Communications.

- Administration, Finance, and Logistics.

- Plan Development and Maintenance.

- Authorities and References.

The basic plan provides the general approach to emergency response, but it does not stand alone. Unit 5 will discuss functional annexes, and hazard-, threat-, and incident-specific annexes or appendices to the EOP.

Knowledge Check

Carefully read each question and all of the possible answers before selecting the most appropriate response for each test item. Circle the letter corresponding to the answer you have chosen, or fill in the blank.

1. Which section of the EOP includes information about hazards?

   a. Purpose statement
   b. Situation and Assumptions
   c. Concept of Operations
   d. Administration and Logistics

2. The _____ section of the EOP lists organizations and their responsibilities in emergency response.

3. Which section of the EOP specifies the sequence of actions before, during, and after an event?

   a. Concept of Operations
   b. Organization and Assignment of Responsibilities
   c. Plan Development and Maintenance
   d. Authorities and References

4. The Administration and Logistics section addresses resource needs and availability.

   a. True
   b. False

5. Which of the following is not described in the Plan Development and Maintenance section?

   a. Planning participants
   b. Planning responsibilities
   c. The revision cycle
   d. Resources

Knowledge Check (Continued)

1. b
2. Organization and Assignment of Responsibilities
3. a
4. a
5. d

Unit 5: Annexes and Appendices

Introduction and Unit Overview

This unit will introduce functional annexes and hazard-, threat-, and incident-specific appendices. After you complete this unit, you should be able to:

- Describe the difference between an annex and an appendix.

- List the functional annexes recommended for inclusion in every EOP.

- Identify factors for deciding whether to develop an annex or an appendix.

- Describe the basic structure for annexes and appendices.

- Determine when other documentation (e.g., maps) is needed to accompany annexes or appendices.

Annexes vs. Appendices—What Is the Difference?

How do you know whether to develop an annex or appendix to your community's EOP? The first step is to understand the difference between an annex and an appendix. Annexes and appendices are different in content and address different topics.

An **annex** explains *how the community will carry out a broad function in any emergency, such as warning or resource management.*

An **appendix** is a *supplement to an annex that adds information about how to carry out the function in the face of a specific hazard.* Thus every annex may have several appendices, each addressing a particular hazard. Which hazard-specific appendices are included depends on the community's hazard analysis. For example, a community in California would probably include earthquake appendices in its EOP; a community in Florida would probably include hurricane appendices; and a community in the Midwest would undoubtedly include appendices that address tornadoes. The decision about whether to develop an appendix rests with the planning team.

In general, the organization of annexes and appendices parallels that of the basic plan. Specific sections can be developed to expand upon—but not to repeat—information that is contained in the basic plan.

Functional Annexes

It is important early in the planning process to choose the functions that will be included in the basic plan as annexes. Factors that you should consider when making these choices include the:

- **Organizational structure** of the State and local governments.

- **Capabilities** of your jurisdiction's emergency services agencies.

- **Established policy** in regard to the concept of operations.

Keep in mind the threat analysis information that you learned in Unit 3. What you and your planning team know about the vulnerability of your community is key to developing meaningful functional annexes to your basic plan.

Because communities vary so widely, no single listing of functional annexes can be prescribed for every one. However, nine core functions are typically addressed in annexes in every EOP:

- **The Direction, Control, and Coordination Annex** allows a jurisdiction to analyze the situation and decide on the best response, direct the response teams, coordinate efforts with other jurisdictions, and make the best use of available resources.

- **The Information Collection and Dissemination Annex** describes the required critical or essential information common to all operations identified during the planning process.

- **The Communications Annex** provides detailed focus on the total communications system and how it will be used.

- **The Population Warning Annex** describes the warning systems in place and the responsibilities and procedures for using them.

- **The Emergency Public Information (EPI) Annex** provides the procedure for giving the public accurate, timely, and useful information and instructions throughout the emergency period.

  Note that while a *warning* annex focuses on the procedures that the *government uses to alert those at risk*, an *EPI* annex deals with *developing messages* and accurate information, *disseminating* the information, and *monitoring* how the *information* is received. Because the warning system is one means for an EPI organization to get information out, an EPI annex must address coordination with those responsible for the warning system.

- **The Public Protection Annex** describes the provisions (e.g., for evacuation or in-place sheltering) that have been made to ensure the safety of people threatened by the hazards the jurisdiction faces.

Functional Annexes (Continued)

- **The Mass Care and Emergency Assistance Annex** deals with the actions that are taken to protect evacuees and other disaster victims from the effects of the disaster, including providing temporary shelter, food, medical care, clothing, and other essential needs.

- **The Health and Medical Services Annex** describes policies and procedures for mobilizing and managing health and medical services under emergency or disaster conditions.

- **The Resource Management Annex** describes the means, organization, and process by which a jurisdiction will find, obtain, allocate, and distribute resources to satisfy needs that are generated by an emergency or disaster.

In addition to the above listed annexes, the EOP planning team should include annexes that make sense for your community. For example, if your community has a nuclear power plant, you may want to include an annex on radiological protection. Other functional annexes that may be included are:

- Damage assessment.

- Search and rescue.

- Emergency services.

- Aviation operations.

Some jurisdictions may want to modify their Functional Annex structure to use the 15 ESFs identified in the NRF. Some communities that have adopted the ESF approach have also added additional ESFs to meet Local needs. The ESF structure facilitates the orderly flow of Local requests for governmental support to the State and Federal levels and the provision of resources back down to local government during an emergency. State and Local jurisdictions that choose not to adopt the ESF structure should cross-reference their Functional Annexes with the ESFs.

Activity: Reviewing Your EOP's Functional Annexes

To complete this activity, refer to your community's basic plan that you reviewed in Unit 4. The questions below will help you organize your review.

Again, while no single listing of annexes can be prescribed for every community, you must be aware of which annexes (if any) your EOP contains and which annexes are required for your jurisdiction based on your resources, size and type of government, etc.

Remember that, to be effective, the annexes in your EOP must include core functions as well as functions that address the hazards that you learned to profile in Unit 3.

1. Does your community's plan include a(n):

   Direction and Control Annex?

   ☐ Yes
   ☐ No

   Communications Annex?

   ☐ Yes
   ☐ No

   Population Warning Annex?

   ☐ Yes
   ☐ No

   Emergency Public Information (EPI) Annex?

   ☐ Yes
   ☐ No

   Public Protection Annex?

   ☐ Yes
   ☐ No

   Mass Care and Emergency Assistance Annex?

   ☐ Yes
   ☐ No

Activity: Reviewing Your EOP's Functional Annexes (Continued)

Health and Medical Services Annex?

☐ Yes
☐ No

Resource Management Annex?

☐ Yes
☐ No

2. What other annexes, if any, does your EOP contain?

3. Based on the knowledge of threat analysis that you gained in Unit 3, what, if any, additional annexes should your EOP contain?

4. What threats are addressed by the plan?

5. Who has the authority to activate part or all of the plan?

Hazard-, Threat-, or Incident-Specific Appendices

Hazard-, threat-, or incident-specific appendices are attached to each functional annex to specify how that function should be carried out in the face of a particular hazard, threat, or incident. Topics addressed in these appendices include:

- Special planning requirements.

- Priorities identified through threat analysis.

- Unique characteristics of the hazard, threat, or incident requiring special attention.

- Special regulatory considerations.

The table below and on the following page suggests appendix topics for each functional annex.

Annex	Appendix Topics
Direction and Control	• Response actions keyed to specific time periods and phases. • Urban Search and Rescue (US&R) inspection. • Inspection, condemnation, and demolition of structures and buildings. • Protective gear for responders. • Detection equipment and techniques. • Laboratory analysis services. • Containment and clean-up teams. • Actions to ensure that the area directly affected by the disaster is secure and safe enough for the return of evacuated populations or for the continued presence of those who did not evacuate.
Communications	• Provisions made to ensure that the effects of a specific hazard do not prevent or impede the ability of response personnel to communicate with each other during response operations.
Population Warning	• Hazard-unique public warning protocols. • Required or recommended notifications of State and Federal officials.
Emergency Public Information	• Information the public will need to know about the specific hazard (e.g., special evacuation routes and shelters, in-place protective actions, etc.). • The means (i.e., particular medium/media) that will be used to convey that information to the public.

Hazard-, Threat-, or Incident-Specific Appendices (Continued)

Annex	Appendix Topics
Public Protection	• Public protection options and timing. • Special exclusion zones for a specific hazard (e.g., downwind and crosswind areas for nuclear power and chemical plants; coastal areas subject to flooding caused by storms, hurricanes, and/or tidal surge, etc.). • Evacuation routes. • Transportation resources to support mass evacuation.
Mass Care and Emergency Assistance	• Shelter locations outside the hazard's vulnerable area. • Structural survivability requirements for building in the hazard vulnerability zone and the application of mitigation measures. • Protection of shelter occupants from the effects of the hazard. • Special medicines and/or antidotes for shelter occupants. • Food and water stocks to support extended shelter stays. • Capability to decontaminate people exposed to hazardous materials.
Health and Medical Services	• Unique health consequences and treatment options for people exposed to the hazard. • Environmental monitoring and/or decontamination requirements.
Resource Management	• Provisions for purchasing, stockpiling, or otherwise obtaining special protective gear, supplies, and equipment needed by response personnel and disaster victims.

Annex and/or Appendix Implementing Instructions

Each annex or appendix (as well as the basic plan) may use implementing instructions in the form of:

- Standard Operating Procedures (SOPs).

- Maps.

- Charts.

- Tables.

- Forms.

- Checklists.

Implementing instructions may be included as attachments or referenced. Your EOP planning team may use supporting documents as needed to clarify the contents of the plan, annex, or appendix. For example, the Evacuation Annex may be made clearer by attaching maps with evacuation routes marked. Because these routes may change depending on the location of the hazard, maps may also be included in the hazard-specific appendices to the Evacuation Annex. Similarly, the locations of shelters may be marked on maps supporting the Mass Care Annex.

An example of the kind of forms that may be attached is a resource request form. Charts, tables, and checklists are appropriate for the Organization and Assignment of Responsibilities Annex (e.g., a matrix of position responsibilities).

Activity: Appendix Review

Refer to your community's basic plan to complete this activity. Choose an appendix to one of your EOP's functional annexes and use the checklist below to review it for completeness.

If your plan contains no appendices, choose a functional annex and use the checklist to complete a preliminary outline of an appendix that would specify how the annex's function should be carried out.

Annex _____**Appendix**_____

Does this appendix address one or more of the following:

- ☐ Unique planning requirements uncommon to other hazards?
- ☐ Regulatory requirements associated with the specific hazard?
- ☐ Hazard-specific information not covered in the annex?

Does this appendix:

- ☐ Pertain to the annex to which it is appended?
- ☐ Provide an appropriate supplement to the annex?
- ☐ Avoid duplication of information in the annex?
- ☐ Quantify:

    - ☐ Risk area?
    - ☐ Geography?
    - ☐ Demographics?
    - ☐ Other?

- ☐ Provide enough information for the jurisdiction to carry out the function?
- ☐ Provide work aids such as maps, charts, and checklists, as appropriate?

Is the organization of this appendix consistent with other parts of the EOP (i.e., Purpose, Situation and Assumptions, etc.)?

- ☐ Yes
- ☐ No

What modifications or additions would you suggest to make this appendix more effective?

Unit Summary

An annex explains how the community will carry out a broad function in any emergency. There are nine core functions that typically are addressed in annexes in every EOP:

- Direction and Control

- Information Collection and Dissemination

- Communications

- Population Warning

- Emergency Public Information (EPI)

- Public Protection

- Mass Care and Emergency Assistance

- Health and Medical Services

- Resource Management

An appendix is a supplement to an annex that adds information about how to carry out the function in the face of a specific hazard, threat, or incident. Thus, every annex may have several appendices.

Each annex or appendix (as well as the basic plan) may use supporting documents in the form of:

- SOPs.

- Maps.

- Charts.

- Tables.

- Forms.

- Checklists.

Each jurisdiction's planning team should assure the inclusion in their EOP of the functional annexes and hazard-specific appendices that will meet the requirements of their community during emergencies and disasters.

Unit 6 will discuss implementing instructions for carrying out the tasks that are assigned in the EOP.

Knowledge Check

Carefully read each question and all of the possible answers before selecting the most appropriate response for each test item. Circle the letter corresponding to the answer you have chosen.

1. An _____ describes a particular disaster response function.

   a. Annex
   b. Appendix

2. An annex is attached to an appendix.

   a. True
   b. False

3. Which of the following is <u>not</u> one of the functional annexes that should be included in every EOP?

   a. Direction and Control
   b. Communications
   c. Aviation Operations
   d. Population Warning

4. Which of the following is <u>not</u> a type of supporting document?

   a. Map
   b. Recordkeeping form
   c. Checklist
   d. References

5. Response protocols for a tornado would be outlined in an:

   a. Annex.
   b. Appendix.

 **Knowledge Check (Continued)**

1. a
2. b
3. c
4. d
5. b

Unit 6: Implementing Instructions

Introduction and Unit Overview

This unit will introduce the different types of implementing instructions that may be developed at the agency level and how they are used. After you complete this unit, you should be able to:

- Describe the different types of implementing instructions and the purpose for each.

- Explain why implementing instructions must be developed at the agency level.

- Identify an implementing instruction that would be appropriate for a tasking or an area of responsibility assigned to your agency.

What Are Implementing Instructions?

Implementing instructions are documents, developed by individual agencies, that provide detailed instructions for carrying out tasks assigned in the EOP. Typically, implementing instructions are included in the EOP by reference only.

Implementing instructions provide tools for carrying out the community's plan. They help ensure that those who are responsible for implementing the EOP are able to carry out their roles effectively.

There are several types of implementing instructions that organizations can develop:

- Standard Operating Procedures (SOPs)

- Job aids

- Checklists

- Information cards

- Recordkeeping and combination forms

- Maps

Each of these types of implementing instructions will be discussed in the sections that follow.

Standard Operating Procedures

SOPs:

- Provide step-by-step instructions for carrying out specific responsibilities.

- Describe who, what, where, when, and how.

- Are appropriate for:

  - Complex tasks requiring step-by-step instructions.
  - Tasks for which standards must be specified.
  - Tasks for which documentation of performance protocols are required as protection against liability.

To develop SOPs:

- Develop a task list.

- Determine who, what, where, when, and how. Note that who includes:

  - Who performs the activity.
  - To whom he or she reports.
  - With whom he or she coordinates.

- Identify the steps for each task.

- Identify the standards for task completion.

- Test the procedures.

Keep the SOPs up-to-date through review and revision.

Job Aids

SOPs, or parts of SOPs, can be presented as job aids. A job aid is a written procedure that is intended to be used *on the job* while the task is being done. Job aids are appropriate for complex tasks, critical tasks that could result in serious consequences, tasks that are infrequently done, and for procedures or personnel that change often. Job aids are also useful when conformity is needed among workers and/or across locations. Job aids should specify:

- The task *title*.

- The *purpose* of the task.

- *When* to do the task.

- *Materials* needed to perform the task.

- *How* to perform each step of the task.

- The desired *result(s)*.

- *Standards* to which the task must be performed.

- How to *check* the work.

A job aid may include:

- Examples.

- Graphics.

- Flow charts.

- If...then decision tables.

- Dos and don'ts.

Because job aids are designed to be used in the midst of completing a task, to be effective they must be clear. They should use action verbs and everyday language, highlight important information, and place warnings *before* the steps to which they apply.

Formatting is also important when creating job aids. Numbering steps and using space, boxes, or lines to separate steps allows users to find their place easily after looking away.

Job Aids (Continued)

You should keep in mind that job aids may not be useful for all tasks, especially simple tasks that are performed regularly, or must be accomplished quickly, from memory. If a task cannot be performed while referring to a job aid at the same time, a job aid is not appropriate for the task.

Checklists

Checklists provide a list of tasks, steps, features, contents, or other items to be checked off. They often take the form of boxes to be checked off (e.g., yes/no, done/not done, present/not present). They may also take the form of a rating scale.

Checklists are useful for tasks that are made up of multiple steps or for when it might be necessary to document the completion of the steps. Checklists are less useful when observations must be recorded or when calculations or evaluations must be made.

Information Cards

Information cards provide information that is needed on the job in a convenient—often graphic—form. Examples include:

- Reference lists.

- Diagrams, labeled illustrations, charts, or tables.

- Information summarized in matrix form (e.g., a tax table).

Information that might be usefully presented on information cards includes:

- Call-down rosters.

- Contact lists.

- Resource lists.

- Organizational charts.

- Task matrices.

- Equipment diagrams.

Forms

Common forms used as implementing instructions include:

- *Recordkeeping* forms on which calculations, observations, or other information (e.g., damage assessment) can be recorded.

- *Combination* forms that serve multiple functions, such as checklists with recordkeeping sections.

Maps

Finally, *maps* may also be used as implementing instructions to highlight:

- Geographic features and boundaries.

- Jurisdictional boundaries.

- Locations of key facilities.

- Transportation or evacuation routes.

When a map is designed to show a particular feature, extraneous details are often eliminated.

Creating Effective Implementing Instructions

To be effective, implementing instructions must be appropriate for both the intended use and audience. They must also be:

- *Complete* in that they cover all of the components or steps.

- *Clear, concise,* and *easy to use.* They should avoid jargon and ambiguity, be organized logically (the way the task is actually done), and include instructions that identify the purpose and applicability of the particular implementing instruction.

- *Sufficiently detailed* in that they give all of the necessary information.

- *Up-to-date.* The latest revision date should be included.

- *Sufficient in scope* in that, together, they cover each function fully.

- *Identified in the EOP* so that their existence is recorded. Implementing instructions can be incorporated or referenced.

The activity that follows asks you to choose the appropriate type of implementing instruction for the stated purpose.

Activity: Which Type Is Best?

For each item in the first column, choose the type of implementing instruction that best fits from the second column. Then turn the page to check your answers.

_____ 1. A task for which it is necessary to document completion of multiple steps

_____ 2. A call-down roster

_____ 3. A complex task

_____ 4. Evacuation routes

_____ 5. An aid to be used on the job

_____ 6. A means to record observations and/or calculations (e.g., damage assessment)

A. SOPs

B. Job aid

C. Checklist

D. Information card

E. Form

F. Map

Activity: Which Type Is Best? (Continued)

1. c
2. d
3. a
4. f
5. b
6. e

Who Uses Implementing Instructions?

Implementing instructions are used by all agency personnel who respond to disasters, whatever their function.

Implementing instructions are developed at the agency level for two reasons:

- Personnel from an agency with a specific function (e.g., communications) will have no idea how to tell another agency's personnel (e.g., firefighters) how to do their jobs.

- Agency personnel will be the persons who use the implementing instructions and therefore will know if they are helpful and effective.

The implementing instructions used by your agency personnel should support your agency's roles and responsibilities as described in the basic plan. Thus, probably only some types of implementing instructions will be useful to your agency, depending on its function in disaster response.

The following activity gives you an opportunity to identify which implementing instructions would be helpful for your agency to develop to carry out its responsibilities in an emergency.

Activity: Identifying Possible Agency Implementing Instructions

Answer the questions below to identify which implementing instructions would be useful to develop for your agency.

1. What are your agency's primary responsibilities in disaster response?

2. Which implementing instructions would be useful for helping personnel carry out those responsibilities?

Unit Summary

Implementing documents, developed by individual agencies, provide agency personnel with detailed instructions for carrying out their tasks, as assigned in the EOP.

Depending on an agency's roles and responsibilities, different types of implementing instructions may be appropriate tools to help personnel carry out various tasks.

Types of implementing instructions include:

- **SOPs**, appropriate for complex tasks that require step-by-step instructions or tasks for which standards must be specified.

- **Job aids** to help complete on-the-job tasks that are complex, infrequently performed, or that require conformity among personnel or locations.

- **Checklists** that provide a list of tasks, steps, or other items to be checked off—useful for tasks with multiple steps or when documentation of completion of steps is required.

- **Information cards** that provide convenient, graphic, on-the-job information such as diagrams, illustrations, tables or charts.

- **Recordkeeping** forms for recording calculations, observations, or information such as damage assessments.

- **Combination forms** that serve multiple functions.

- **Maps** that indicate information such as geographic features, boundaries, key facilities, evacuation routes, etc.

Unit 7 will review the course content and help you prepare to take the final exam.

Knowledge Check

Carefully read each question and all of the possible answers before selecting the most appropriate response for each test item. Circle the letter corresponding to the answer you have chosen, or fill in the blank.

1. Some SOPs can be presented as job aids to be used in the midst of completing a task.

   a. True
   b. False

2. A job aid provides convenient, sometimes graphic, information that is most useful when completing simple tasks that can be done quickly from memory.

   a. True
   b. False

3. A(n)_____ would be the best type of implementing instruction for a contact list.

4. Implementing instructions are developed at the _____ level.

   a. Federal
   b. State
   c. Agency

5. Implementing instructions must be attached directly to the EOP, not simply incorporated by reference.

   a. True
   b. False

Knowledge Check (Continued)

1. a
2. b
3. Information card
4. c
5. b

Unit 7: Course Summary

Introduction

This course was designed to provide you with training in the fundamentals of the emergency planning process. This unit will review the important information from the entire course and serve as preparation for the final examination. After you complete this unit, you should be able to summarize the key points of the course and complete the final exam.

The Planning Process

Emergency planning is a continual cycle of planning, training, exercising, and revision that takes place throughout the five phases of the emergency management cycle: Prevention, preparedness, mitigation, response, and recovery.

The end product of emergency planning is a community EOP. An **EOP** can be defined as *a document describing how citizens and property will be protected in a disaster or emergency.*

There are six steps in the emergency planning process:

1. **Form a collaborative planning team.**

2. **Understand the situation.**

3. **Determine goals and objectives.**

4. **Develop the plan.**

5. **Prepare, review, and approve the plan.**

6. **Refine and execute the plan.**

Emergency planning necessitates a coordinated team effort involving different levels of government and many community agencies and organizations.

The Planning Process (Continued)

Working with personnel from other agencies and organizations requires collaboration. **Collaboration** is the *process in which people work together as a team on a common mission*—in this case, development of a community EOP. Collaboration is more difficult, however, in the face of differences among agencies in terminology, experience, mission, and culture. Thus, collaboration also requires flexibility to agree on common terms and priorities, and humility to learn from others' ways of doing things.

Collaboration does not come automatically. Building a team that works well together takes time and effort.

Teams go through stages in the process of becoming productive.

- Forming

- Storming

- Norming

- Performing

- Adjourning

The team leader is important in motivating members to work through their differences and focus on the common goal of developing an EOP. Other team members may also play different roles, including task master, innovator, organizer, evaluator, and finisher.

A productive team displays the following characteristics:

- Commitment to a common goal (i.e., development of an EOP).

- A leader who provides direction and guidance.

- Open communication.

- Constructive conflict resolution.

- Mutual trust.

- Respect for each individual and his or her contributions.

Threat Analysis

Threats are *indications of possible violence, harm, or danger.*

Threat analysis determines:

- What can occur.

- How often it is likely to occur.

- The damage it is likely to cause.

- How it is likely to affect the community.

- How vulnerable the community is to the threat.

The steps in the threat analysis process are:

1. **Identify threats.** Develop a list of threats the community may face based on historical data about past events. If your community has an existing threat analysis, review it and identify any changes that have occurred since it was developed.

2. **Profile threats,** considering the duration, seasonal pattern, and speed of onset. A crucial part of the threat profile is the availability of warnings.

3. **Develop a community profile,** considering geography, property, infrastructure, demographics, and response organizations.

4. **Determine vulnerability** by merging information from the community and hazard profiles to focus on the threats that present the highest risk.

5. **Create and apply scenarios** based on the community's threat and risk data.

Threat Analysis (Continued)

Risk is the *predicted impact that a hazard would have on the people, services, and specific facilities in the community.* Quantifying risk involves:

- Identifying the elements of the community that are potentially at risk from a specific threat.

- Developing response priorities.

- Assigning severity ratings.

- Compiling risk data into community risk profiles.

In surveying risk, it is helpful to develop response priorities. The following is a suggested hierarchy for setting priorities:

1. Life safety.

2. Essential facilities.

3. Infrastructure lifelines.

The final step in the hazard analysis *process is creating and applying scenarios* based on the community's threat and risk data. A scenario should describe the:

- Initial warning of the event.

- Potential impact on the community.

- Potential impact on specific community sectors.

- Potential consequences.

- Actions and resources needed to deal with the situation.

The Basic Plan

The format of the EOP includes three parts:

- The basic plan.

- Functional annexes.

- Hazard-, threat-, or incident-specific appendices.

The recommended format for the basic plan includes the following components:

- **Introduction,** which includes:

  - The Promulgation document, signed by the jurisdiction's CEO.
  - Signature page, signed by all response organizations.
  - Dated title page and record of changes to the EOP.
  - Record of distribution, which lists EOP recipients.
  - Table of contents.

- **Purpose Statement,** which includes the EOP's purpose and a synopsis.

- **Scope,** which states the scope of emergency and disaster response to which the plan applies and the entities and geographic areas to which it applies.

- **Situation and Assumptions.** The Situation characterizes the community, including hazards and populations. The Assumptions Statement delineates what was assumed to be true when the EOP was developed.

- **Concept of Operations,** which explains the community's approach to emergency response, including activation levels and the sequence of actions before, during, and after an emergency.

- **Direction, Control, and Coordination,** which describes the framework for all direction, control, and coordination activities.

- **Information Collection and Dissemination,** which describes the required critical or essential information common to all operations identified.

- **Communications,** which describes communication protocols and coordination procedures.

The Basic Plan (Continued)

- **Organization and Assignment of Responsibilities,** which assigns responsibilities by organization and position and identifies shared responsibilities (specifying who has primary responsibility and who plays a support role).

- **Administration and Logistics,** which addresses resources.

- **Plan Development and Maintenance,** which assigns planning responsibilities to participants and describes the revision cycle.

- **Authorities and References,** which cites the legal basis for emergency operations and relevant reference materials.

Annexes and Appendices

Annexes and appendices are different in content and address different topics.

An **annex** explains *how the community will carry out a broad function in any emergency, such as communications or evacuation.*

An **appendix** *is a supplement to an annex that adds information about how to carry out the function in the face of a specific hazard.* Thus, every annex may have several appendices, each addressing a particular hazard. Which hazard-specific appendices are included depends on the community's hazard analysis.

Both annexes and appendices are organized in similar fashion to the basic plan.

There are eight functions that typically are addressed in annexes in every EOP (one function per annex):

- Direction, Control, and Coordination

- Information Collection and Dissemination

- Communications

- Population Warning

- Emergency Public Information

Annexes and Appendices

- Public Protection

- Mass Care

- Health and Medical Services

- Resource Management

In addition, communities may choose to add other annexes that make sense for their situations (e.g., radiological protection if the community has a nuclear power plant).

Hazard-specific appendices are attached to each functional annex to specify how that function should be carried out in the face of a particular hazard. Topics addressed in hazard-specific appendices include special regulatory considerations, planning requirements, priorities, and unique hazard characteristics that require special planning.

In addition, each annex or appendix may use supporting documents as needed to clarify the contents (e.g., a map to show evacuation routes, a form to request resources, or a table to list position responsibilities).

Implementing Instructions

Implementing instructions are documents, developed by individual agencies, that provide detailed instructions for carrying out tasks assigned in the EOP.

There are several types of implementing instructions that organizations can develop:

- **SOPs,** which provide step-by-step instructions for carrying out specific responsibilities.

- **Job aids,** which are written procedures that are intended to be used *on the job* while the task is being done.

- **Checklists,** which provide a list of tasks, steps, contents, or other items to be checked off.

- **Information cards,** which provide information such as lists of contacts or resources in a convenient form. Information cards sometimes use graphics such as diagrams.

- **Recordkeeping and combination forms** on which to record calculations and observations.

- **Maps** on which to mark boundaries, key facilities, and routes.

Final Steps

You have now completed IS 235 and should be ready to take the final exam.

Complete the final exam in the back of the book by marking the correct responses.

To submit the final exam online, go to http://training.fema.gov/IS and click on the courses link. Click on the title for this course, and scroll down the course description page to locate the final exam link. After you have selected the final exam link and the online answer sheet is open, transfer your answers, and complete the personal identification data requested.

Appendix A: Sample Plan: Jefferson County

Record Of Changes

The Emergency Management Agency (EMA) ensures that necessary changes and revisions to plan are prepared, coordinated, published and distributed.

The plan will undergo revision whenever:

- Any other condition occurs that causes conditions to change.

- It fails during emergency.

- Exercises, drills reveal deficiencies or "shortfall(s)."

- Local government structure changes.

- Community situations change.

- State requirements change.

EMA will maintain a list of individuals and organizations which have controlled copies of the plan. Only those with controlled copies will automatically be provided updates and revisions. Plan holders are expected to post and record these changes. Revised copies will be dated and marked to show where changes have been made.

"Record of Changes" Form is on the following page.

Record Of Changes

Nature of Change	Date of Change	Page(s) Affected	Changes Made by (Signature)

How To Use This Plan

Note: This is a generic, strategic plan, organized by "emergency functions." The "Basic Plan" section provides a general overview and summary of the purpose, responsibilities, and operational concepts. The schedule of Annexes are functions that may be activated and performed during emergencies and disasters. While the concept of operations should always remain the same, the functions activated will be dependent on the emergency/disaster type and scope.

1. Read the "Basic Plan," Section 1. Take note of your department's/agencies' general responsibilities contained within the "Basic Plan."

2. Look at the "Primary/Support Matrix" appended to the Basic Plan (Section 1: VI. Attachments-A). Find the name of your agency/organization. Note which Annex(es) your department/agency appears in. Within each Annex that your agency is a part of, you will find additional specific responsibilities.

3. Your department/agency must develop and maintain "Standard Operating Procedures (SOPs)" in such detail as necessary that will result in successful activation and completion of your responsibilities as stated. Refer to Section 1: II Organization and Responsibilities Part B of the Basic Plan for additional information and guidance.

   Helpful Hint: Make a list of your general responsibilities (found in Section 1: II Organization and Responsibilities Part B of the Basic Plan) and specific responsibilities (found in each Annex that your department/agency is involved with). This responsibilities listing is the basis for internal, tactical SOPs and personnel action guides.

4. Each Annex contains a "cover page summary." This summary lists the departments/agencies/organizations providing primary support to this function. The county, State, and Federal agencies, who would provide additional assistance if requested, are also shown.

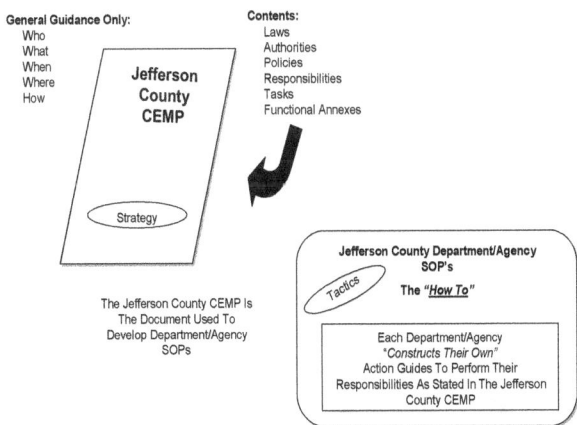

Jefferson County, Alabama
Comprehensive Emergency Management Plan

Section 1: Basic Plan

I. **Mission**
   A. Purpose, Goals, and Definitions
   B. Authorities, Guidance Documents, and Mutual Aid Agreements
   C. Situations
   D. Assumptions
   E. Limitations
   F. Policies

II. **Organization and Responsibilities**
   A. Organization
   B. Responsibilities

III. **Concept of Operations**
   A. General
   B. Emergency Management Phases
   C. Functional Annex Concept

IV. **Administration and Logistics**
   A. General
   B. Policies
   C. Administration
   D. Fiscal
   E. Logistics
   F. Insurance

V. **Direction and Control**
   A. General
   B. Crisis Monitoring
   C. Levels of Emergency
   D. On-scene Management/ICS
   E. EOC Activations and Staffing
   F. Controls, Continuity of Operations
   G. Facilities
   H. Military Support
   I. Continuity of Government
   J. Plan Maintenance

VI. **Attachments**
   A. Primary/Secondary Matrix
   B. Job Aids
   C. List of Acronyms

Jefferson County, Alabama
Comprehensive Emergency Management Plan

Section 1: Back Plan

I. Mission

A. Purpose, Goals, and Definitions

1. Purpose

a. This plan describes the basic strategies, assumptions and mechanisms through which the departments/agencies within the county will mobilize resources and conduct activities to guide and support local emergency management efforts through response and recovery. To facilitate effective intergovernmental operations, this plan adopts a functional approach that groups the type of assistance to be provided under annexes to address functional needs at the local government level. Each "functional annex" is headed by a primary agency, which has been selected based on its authorities, resources, and capabilities in the functional area. The "Functional Annexes" serve as the primary mechanism through which assistance is managed in an affected area.

b. This plan provides for an orderly means to prevent or minimize (mitigation strategies), prepare for, respond to, and recover from emergencies or disasters that threaten life, property, and the environment within Jefferson County boundaries by:

- Identifying major natural and manmade hazards, threats to life, property, and/or the environment that are known or thought to exist.

- Assigning emergency management responsibilities and tasks.

- Describing predetermined actions (responsibilities, tasks) to be taken by departments/agencies, and other cooperating organizations and institutions, to eliminate or mitigate the effects of these threats, and to respond effectively and recover from an emergency or disaster.

- Providing for effective assignment and utilization of local government employees.

- Documenting the current capabilities and existing resources of departments/agencies and other cooperating organizations and institutions which must be maintained to enable accomplishment of those predetermined actions.

- Providing for the continuity of local government during and after an emergency or disaster.

- Enhance cooperation (mutual aid agreements and memorandums of understanding) and coordination with cooperating community agencies, neighboring jurisdictions, and State and Federal agencies.

- Providing for an emergency planning team comprised of representatives from all departments as identified and utilized through this plan development for containing review and revision of the plan; exercise planning and evaluation; and reviewing and offering recommendations on emergency management initiatives.

c. This plan provides guidance for:

- Mitigation, preparedness, response, and recovery policy and procedures.

- Disaster and emergency responsibilities.

- Training and public education activities.

d. This plan is strategic and "responsibility oriented," and addresses:

- Coordinated regional and interregional evacuation, shelter, and post-disaster response and recovery.

- Rapid deployment and predeployment of resources.

- Communications and warning systems.

- Annual exercises to determine the ability to respond to emergencies.

- Clearly defined responsibilities for departments/agencies through a "Functional Annex" approach to planning and operations.

2. Goals

a. Develop citizen self-sufficiency.

b. Develop first responder capabilities.

c. To have a plan (framework, strategy) that will guide organizational behavior (response) during emergency(ies) or disaster(s).

d. Create a framework of interagency and community-wide cooperation to enhance disaster mitigation, preparedness, response, and recovery.

3. Definitions

a. The term "emergency" as used in this plan means a set of circumstances which demand immediate action to protect life, preserve public safety, health and essential services, or protect property and the environment.

b. "Disaster" means the situation requires all available local government resources and/or augmentation, and is beyond the capabilities of the county or a city(ies). A state of "emergency" or "disaster" can be proclaimed by a chief elected official and/or Emergency Council Chairperson, Vice Chairperson, or Emergency Management Agency (EMA) Coordinator.

B. Authorities, Guidance Documents, Mutual Aid Agreements

1. Authorities

a. Federal

- Public Law (P.L.) 93-288, Disaster Relief Act of 1974, as amended by P.L. 100-707 ("The Stafford Act")
- Emergency Management and Assistance, 44 U.S. Code 2.1 (Oct. 1, 1980)
- P.L. 81-920, Federal Civil Defense Act of 1950 as amended
- P.L. 99-499, Title III, Emergency Planning and Community Right-to-Know,   Oct. 17, 1986

b. State

- Public Law (P.L.) 31-9, Act 47, Alabama Civil Defense Act of 1955
- P.L. 29-3, Act 875, Emergency Interim Succession Act, 1961
- P.L. 83-612, Alabama Department of Civil Defense name changed to Alabama Emergency Management Agency, 1983
- Alabama Executive Order Number 4, March 6, 1987
- Alabama Executive Order Number 40, July 23, 1985

c. Local

- Jefferson County Commission Resolution of November 13, 1951
- City of Birmingham General Ordinances, 1980, sections 2-3-1 and 9-4-1 through 9-4-4
- Birmingham/Jefferson County Emergency Management Agency Mayor's Council Resolution of October 10, 1984

2. Guidance Documents

a. State of Alabama Guide for Developing an Emergency Operations Plan, 1996

b. Federal Response Plan, 1992

3. Agreements and Understandings

Central Alabama Fire Chief's Association Disaster Assistance Agreement

C. Situations

1. Hazard Analysis

a. The Jefferson County Hazard Vulnerability Analysis (published separately) provides details on local hazards to include type, effects, impacts, risk, capabilities, and other related data.

---

b. Due to its location and geological features, Jefferson County is vulnerable to the damaging affects of certain hazards that include, but are not limited to:

Natural: Drought, extreme cold, extreme heat, forest fire, riverine flood, flash flood, landshift (earthquake, earthslide, erosion, subsidence), snow/ice/hail, windstorm, lightning storm, hurricane, tornado, epidemic (human/animal).

Technological: Hazardous materials (fixed facility, transportation), fire/explosion, building/structure collapse, dam/levee failure, power/utility outage, extreme air pollution, transportation accident (rail, marine, aircraft, motor vehicle).

Civil/Political Disorder: Economic emergency, riot, strike, demonstration/special events, terrorism/sabotage, hostage situation, attack (conventional, nuclear, biological, chemical).

c. Disaster response efforts are often hampered by equipment and facility damage, communication failures, inclement weather, responder injury and death, and many other limiting factors. In the event of an emergency or disaster that exceeds the available resources, the public should expect and be prepared for a minimum 72 hour delay for emergency response service.

D. Assumptions

1. Governmental officials within the county recognize their responsibilities regarding the safety and well being of the public and they will assume their responsibilities when the Comprehensive Emergency Management Plan is implemented.

2. General Conditions. When a community experiences a disaster, its surviving citizens fall into three broad categories: Those directly affected through personal or family injury or property damage; those indirectly affected by an interruption of the supply of basic needs; and those that are not personally impacted. These guidelines were designed to promote citizen self-confidence and independence in the face of a disaster. Following these guidelines will allow the emergency organization within the county to concentrate first on helping those citizens directly affected by a disaster.

3. It is expected that each individual or head of a household will develop a family disaster plan and maintain the essential supplies to be self-sufficient for a minimum of 72 hours.

4. Businesses are expected to develop internal disaster plans that will integrate and be compatible with local government resources and this plan.

Note: This plan is not intended to limit or restrict initiative, judgment, or independent action required to provide appropriate and effective emergency and disaster mitigation, preparation, response, and recovery.

E. Limitations

1. It is the policy of the Jefferson County Emergency Management Council that no guarantee is implied by this plan. Because local government assets and systems may be damaged, destroyed, or overwhelmed, the Council can only endeavor to make responsible efforts to respond based on the situation, information, and resources available at the time.

2. Adequate funding is needed to support this plan and its programs. The performance of the assigned tasks and responsibilities will be dependent on appropriations and funding to support the plan. Lack of funding may degrade the services envisioned under this plan.

   Note: The inability of Departments/Agencies to carry out their responsibilities as indicated in both the Basic Plan and Annexes due to lack of staff and funding may lower "emergency declaration threshold."

F. Policies

1. In order to protect lives and property and in cooperation with other elements of the community (e.g., business, volunteer sector, social organizations, etc.), it is the policy of the Jefferson County Emergency Management Council to endeavor to mitigate, prepare for, respond to, and recover from all natural and manmade emergencies and disasters.

2. It is the policy of the Jefferson County Emergency Management Council that it will take appropriate action in accordance with this plan to mitigate any harm to the citizens or property in the county.

3. Because of the nature of emergencies and disasters (causing damages, interruptions and shortfalls to local government resources), it is the policy of the Jefferson County Emergency Management Council that citizens are encouraged to be self-sufficient for a minimum of 72 hours should an emergency or disaster occur.

4. It is the policy of the Jefferson County Emergency Management Council to make this plan a "user friendly" document.

5. Non-discrimination. It is the policy of the Jefferson County Emergency Management Council that no services will be denied on the basis of race, color, national origin, religion, sex, age, or inability, and no special treatment will be extended to any person or group in an emergency or disaster over and above what normally would be expected in the way of local government services. Council activities pursuant to the Federal/State Agreement for major disaster recovery will be carried out in accordance with Title 44, Code of Federal Regulations (CFR), Section 205.16—nondiscrimination. Federal disaster assistance is conditional on full compliance with this rule.

Jefferson County, Alabama
Comprehensive Emergency Management Plan

Section 1: Back Plan

II. Organization and Responsibilities

A. Organization

1. Emergency Management Council

a. The Jefferson County Commission by November 13, 1951, resolution (pursuant to Federal and State law) created the Birmingham/Jefferson County Civil Defense Corps. The local governing bodies within the county passed resolutions/ordinances joining in this organization creating a "Civil Defense Council" to govern the joint "civil defense program." The Council is authorized and empowered to make, amend, and rescind any and all necessary orders, rules, and regulations for direction and control of the "program." As needed the Council requests municipalities to adopt proper ordinances implementing within each municipality the orders rules, and regulations of the council.

b. As per the 1951 resolution and current state law, the Council is charged with establishing and maintaining an emergency management organization, and developing policies to prepare for, respond to, and recover from emergencies and disasters that threaten or occur in Jefferson County. The policies will be established through the promulgation of a comprehensive Emergency Management Plan.

c. The Council is governed by a Chairperson and Vice Chairperson. In the absence of the Chairperson and the Vice Chairperson, the EMA Coordinator has the responsibility to carry out Council policy in all matters.

d. The Jefferson County Emergency Management Council has designated the EMA Coordinator as being responsible for day-to-day operations, including the implementation of policies and procedures issued by the Council. The EMA Coordinator reports to the Emergency Management Council Chairperson.

2. Emergency Organization

a. The Chief Elected Official of the municipalities, together with the President of the Jefferson County Commission, constitute the Jefferson County Emergency Management Council. The Council is, and has been from its inception, the body charged with the overall responsibility for developing and implementing comprehensive emergency management policies in Jefferson County, Alabama.

b. The Jefferson County Emergency Management Council has selected two members to serve as General Chairperson and Vice Chairperson and to represent the Council in the conduct of emergency operations.

    c. The Primary/Support Matrix (see Basic Plan Attachment A) reflects the organizational structure of the Jefferson County Emergency Management network and indicates the various activities which can support emergency operations.

    d. This plan establishes the emergency management organization within the county. All officers and employees of local government are part of the emergency organization. All appointments and work assignments in an emergency situation shall be documented. All departments/agencies will submit documentation as to staffing allocation, equipment distribution, and other emergency related needs as requested by the EMA Coordinator.

3. Incident Command System

This plan formalizes the Incident Command organization and structure at incident sites.

4. EMA Planning Team

An EMA Planning Team is hereby established through promulgation of this plan. The team shall be composed of representatives or alternates from selected departments/agencies. The team shall formulate emergency management recommendations to the EMA Coordinator. This includes development and maintenance of this plan, exercise planning and evaluation, and related initiatives. Members of the planning team shall be appointed by the EMA Coordinator. The EMA Coordinator will chair the team and will schedule periodic meetings as needed.

B. Responsibilities

1. General Preparedness Responsibilities (All departments/agencies within Jefferson County). The following common responsibilities are assigned to each department/agency listed in this plan. Further, each department/agency shall create an internal emergency management organization and develop Standard Operating Procedures (SOPs) in accordance with the provisions of this plan. Preparation activities include:

- Establishing departmental and individual responsibilities (as indicated in this plan); identify emergency tasks.

- Working with other departments/agencies to enhance cooperation and coordination, and eliminate redundancy. Departments having shared responsibilities should complement each other.

- Establishing education and training programs so that each division, section, and employee will know exactly where, when, and how to respond.

- Developing site specific plans for department facilities as necessary.

- Ensuring that employee job descriptions reflect their emergency duties.

- Training staff and volunteer augmentees to perform emergency duties, tasks.

- Identifying, categorizing and inventorying all available departmental resources.

- Developing procedures for mobilizing and employing additional resources.

- Ensuring communication capabilities with the EOC.

- Filling positions in the emergency organization as requested by the EMA Coordinator acting in accordance with this plan.

- Preparing to provide internal logistics support to department operations during the initial emergency response phase.

2. **General Response Responsibilities** (All departments/agencies within Jefferson County). The following common responsibilities are assigned to each department listed in this plan.

- Upon receipt of an alert or warning, initiate notification actions to alert employees and volunteer augmentees assigned response duties.

- As appropriate:
  - Suspend or curtail normal business activities.
  - Recall essential off-duty employees.
  - Send nonessential employees home.
  - Evacuate departmental facilities.

- As requested, augment the EOC's effort to warn the public through use of vehicles equipped with public address systems, sirens, employees going from door to door, etc.

- Keep the EOC informed of field activities, and maintain a communications link to the EOC.

- Activate a control center to support and facilitate department response activities, maintain events log, and report information to the EOC.

- Report damages and status of critical facilities to the EOC.

- If appropriate or requested, send a representative to the EOC.

  - Ensure staff members tasked to work in the EOC have the authority to commit resources and set policies.

- Coordinate with the EOC to establish protocols for interfacing with State, Federal responders.

- Coordinate with the EOC Information Officer before releasing information to the media.

- Submit reports to the EOC detailing departmental emergency expenditures and obligations.

3. Specific Local Governmental Responsibilities

a. Ambulance Services

- Perform functions in the EOC or on-scene as assigned.
- Provide EMA and/or EOC initial situation/damage reports as per field units' observations and reports from the general public.
- Provide supplies, equipment, and personnel as requested.
- Provide emergency medical transportation and emergency medical services in the field.

b. Birmingham Communications Department

- Perform functions in the EOC or on-scene as assigned.
- Provide EMA and/or EOC initial situation/damage reports as per field units' observations and report from the general public.
- Provide supplies, equipment, and personnel as requested.
- Maintain communication systems.

c. Birmingham Equipment Management

- Perform functions in the EOC or on-scene as assigned.
- Provide EMA and/or EOC initial situation/damage reports as per field units' observations and reports from the general public.
- Provide supplies, equipment, and personnel as requested.
- Operate fleet repair facility.
- Provide for availability of motor fuels for all county vehicles, American Red Cross vehicles, and fuel driven facility.
- Provide for storage of equipment and vehicles in a safe place.
- Provide for security of county complex.
- Provide comprehensive list of county vehicles and equipment to EMA.

d. Birmingham Regional Emergency Medical Services System (BREMSS)

- Perform functions in the EOC or on-scene as assigned.
- Provide EMA and/or EOC initial situation/damage reports as per field units' observations and reports from the general public.
- Provide supplies, equipment, and personnel as requested.
- Provide training and equipment.
- Develop and coordinate field medical protocols.
- Provide trauma coordination through the Trauma Control Center.
- Assist in mass fatality incidents.

e. Business and Industry

- Perform functions in the EOC or on-scene as assigned.
- Provide EMA and/or EOC initial situation/damage reports as per field units' observations and reports from the general public.
- Provide supplies, equipment, and personnel as requested.

f. Building Inspection Services

- Perform functions in the EOC or on-scene as assigned.
- Provide EMA and/or EOC initial situation/damage reports as per field units' observations and reports from the general public.
- Provide supplies, equipment, and personnel as requested.

g. Campuses, Universities

- Perform functions in the EOC or on-scene as assigned.
- Provide EMA and/or EOC initial situation/damage reports as per field units' observations and reports from the general public.
- Provide supplies, equipment, and personnel as requested.
- Provide the estimated number of people requiring emergency shelter, food, and water distribution points, check car facilities as need.

h. Churches

- Perform functions in the EOC or on-scene as assigned.
- Provide EMA and/or EOC initial situation/damage reports as per field units' observations and reports from the general public.
- Provide supplies, equipment, and personnel as requested.
- Provide facilities for emergency shelter, food, and water distribution points, check car facilities as need.

i. Civil Air Patrol

- Perform functions in the EOC or on-scene as assigned.
- Provide EMA and/or EOC initial situation/damage reports as per field units' observations and reports from the general public.
- Provide supplies, equipment, and personnel as requested.
- Augment search and rescue missions as requested and within their capabilities.

j. Community Service Organizations

- Perform functions in the EOC or on-scene as assigned.
- Provide EMA and/or EOC initial situation/damage reports as per field units' observations and reports from the general public.
- Provide supplies, equipment, and personnel as requested.
- Assist with meeting the needs of special populations and individuals.

k. Data Processing (Jefferson County Information Services)

- Perform functions in the EOC or on-scene as assigned.
- Provide EMA and/or EOC initial situation/damage reports as per field units' observations and reports from the general public.
- Provide supplies, equipment, and personnel as requested.
- Maintain computers for payroll.

l.   Department of Human Resources (DHR)

  - Perform functions in the EOC or on-scene as assigned.
  - Provide EMA and/or EOC initial situation/damage reports as per field units' observations and reports from the general public.
  - Provide supplies, equipment, and personnel as requested.
  - Assist with the assessment of human needs during and after a disaster.
  - Identify requirements for individuals and populations with "special needs."
  - Coordinate transportation requirements for special needs agencies/individuals.
  - Coordinate with the American Red Cross, and other agencies as necessary to provide emergency programs for basic human needs to include reception centers, shelters, mass feeding.
  - Work in close concert with American Red Cross and others in the activation and operation of short-term temporary "holding centers" and long-term shelters/disaster centers.

m. Emergency Management Council

  - Provide overall direction and control.
  - Proclaim a county-wide "state of emergency," when necessary.
  - Make disaster declarations and request State and Federal assistance.
  - Issue emergency rules and proclamations that have the force of law during the proclaimed emergency period.
  - Ensure that the county continues to function administratively and make administrative policy decisions.
  - Appropriate funds to meet disaster expenditure needs.
  - Extend or terminate emergency/disaster declarations.

n. Finance

  - Perform functions in the EOC or on-scene as assigned.
  - Provide EMA and/or EOC initial situation/damage reports as per field units' observations and reports from the general public.
  - Provide supplies, equipment, and personnel as requested.
  - Provide appraisers to assist with damage assessments.
  - Process emergency purchases/procurement.
  - Establish and maintain a system whereby incident costs are identified and accumulated for State and Federal reimbursement.
  - Ensure payroll system setup to pay employees.

o. Fire Departments

- Perform functions in the EOC or on-scene as assigned.
- Provide EMA and/or EOC initial situation/damage reports as per field units' observations and reports from the general public.
- Provide supplies, equipment, and personnel as requested.
- Provide, coordinate fire and rescue services.
- Provide initial emergency medical services and prehospital care.
- Contain, control hazardous materials.
- Provide limited response to search and rescue off-road situations, and coordinate heavy rescue operations.
- Augment warning system by providing siren-equipped and/or public address mobile units, and/or manpower for door-to-door warning.
- Support other public safety operations.
- Order evacuation whenever necessary to protect lives and property.

p. Funeral Directors Association

- Assist in mass fatality incidents by providing recovery, evacuation, mortuary operations, identification and notification.

q. Greater Birmingham Humane Society

- Provide assistance in the prevention, detection and control of rabies.
- Provide animal control and services assistance.

r. Hospitals

- Provide medical care.
- Resupply field units with consumable medical supplies.
- Make assessment of hospital capabilities and damages.
- Mobilize staff to provide teams to respond to mass casualty incidents.
- Coordinate with Blood Bank and assist in blood procurement for community needs.
- Participate in hospital radio net that links hospitals, Emergency Operations Center, fire dispatch, and FD/EMS.

s. Jefferson County Communications

- Perform functions in the EOC or on-scene as assigned.
- Provide EMA and/or EOC initial situation/damage reports as per field units' observations and reports from the general public.
- Provide supplies, equipment, and personnel as requested.
- Maintain 800mhz radio system.

t. Jefferson County Coroner/Medical Examiner's Office

- Perform functions in the EOC or on-scene as assigned.
- Provide EMA and/or EOC initial situation/damage reports as per field units' observations and reports from the general public.
- Provide supplies, equipment, and personnel as requested.
- Establish fatality collection areas to facilitate recovery operations.
- Identify mass burial sites.
- Protect the property and personal effects of the deceased.
- Notify relatives.
- Establish and maintain a comprehensive recordkeeping system for continual updating and recording of fatality numbers.

u. Jefferson County Department of Health

- Perform functions in the EOC or on-scene as assigned.
- Provide EMA and/or EOC initial situation/damage reports as per field units' observations and reports from the general public.
- Provide supplies, equipment, and personnel as requested.
- Perform disease control operations, to include epidemic intelligence, evaluation, prevention (including mass inoculations) and detection of communicable diseases.
- Issue general health advisories, information, and instructions.
- Conduct environmental health activities in regard to waste disposal, refuse, food, water control, and vector control.

v. Jefferson County Emergency Management Agency

- Designated by the Emergency council to develop and maintain county-wide emergency management program.
- Develop and maintain the county-wide Comprehensive Emergency Management Plan (CEMP).
- Provide coordination among local, State, Federal, private, and volunteer organizations.
- Maintain liaison with neighboring jurisdictions.
- Develop, coordinate, and maintain warning and emergency communication systems.
- Disseminate emergency alerts and warnings to key officials, department/agencies and the public.
- Disseminate emergency information and instructions.
- Develop, maintain, and disseminate emergency preparedness education materials to include hazard awareness programs.
- Schedule tests, exercises.
- Maintain inventories of resources and equipment.
- Develop mutual aid agreements.
- Monitor situations for CAT or EOC activation.
- Ensure a functional county EOC.
- Coordinate requests for emergency assistance.

- Coordinate and use all available resources during an emergency or disaster.
- Develop and maintain a county-wide damage assessment program, and coordinate assessment procedures with local government.

w. Jefferson County Fleet Services

- Perform functions in the EOC or on-scene as assigned.
- Provide EMA and/or EOC initial situation/damage reports as per field units' observations and reports from the general public.
- Provide supplies, equipment, and personnel as requested.
- Operate fleet repair facility.
- Provide for availability of motor fuels.
- Provide for storage of equipment and vehicles.

x. Jefferson County General Services

- Perform functions in the EOC or on-scene as assigned.
- Provide EMA and/or EOC initial situation/damage reports as per field units' observations and reports from the general public.
- Provide supplies, equipment, and personnel as requested.
- Manage and coordinate volunteers and donations through the EOC.

y. Law Enforcement (Agencies)

- Perform functions in the EOC or on-scene as assigned.
- Provide EMA and/or EOC initial situation/damage reports as per field units' observations and reports from the general public.
- Provide supplies, equipment, and personnel as requested.
- Augment warning system by providing siren-equipped and/or public address mobile units, and/or manpower for door-to-door warning.
- Responsible for lost person search and rescue, and coordination of heavy rescue operations.
- Maintain law and order and provide public safety activities as required.
- Provide security for key facilities.
- Protect property in evacuated areas.
- Provide assistance in the capture and control of animals.
- Enforce orders of fire officers and implement/enforce evacuation orders, when necessary.
- Provide law enforcement and traffic control in support of fire department actions.
- Order/conduct evacuations when necessary to save lives and property.

z. Law Enforcement (Campuses)

- Perform functions in the EOC or on-scene as assigned.
- Provide EMA and/or EOC initial situation/damage reports as per field units' observations and reports from the general public.
- Provide supplies, equipment, and personnel as requested.
- Assist law enforcement services on campuses.

aa. Law Enforcement (Explorers)

- Perform functions in the EOC or on-scene as assigned.
- Provide EMA and/or EOC initial situation/damage reports as per field units' observations and reports from the general public.
- Provide supplies, equipment, and personnel as requested.
- Assist law enforcement agencies within their capabilities.

bb. Law Enforcement (Reserves)

- Perform functions in the EOC or on-scene as assigned.
- Provide EMA and/or EOC initial situation/damage reports as per field units' observations and reports from the general public.
- Provide supplies, equipment, and personnel as requested.
- Assist law enforcement agencies within their capabilities.

cc. Management Information Services

- Perform functions in the EOC or on-scene as assigned.
- Provide EMA and/or EOC initial situation/damage reports as per field units' observations and reports from the general public.
- Provide supplies, equipment, and personnel as requested.
- Provide for the use of computer resources to record and maintain emergency information, data on the organization and operation of congregate care facilities (shelter/mass feeding), and registration of displaced persons.
- Provide assistance in the registration of people at congregate care facilities (shelter/mass feeding).

dd. Media

- In general, disseminate warning messages provided by authorized sources to the general public as rapidly as possible in the event of an impending or actual disaster.
- Commercial Print Media: Assist with emergency information dissemination.
- Commercial Radio and Television Systems: Assist with emergency information dissemination.
- Commercial Radio and Television Systems: Maintain Emergency Alert System.

ee. Metro Area Express

- Perform functions in the EOC or on-scene as assigned.
- Provide EMA and/or EOC initial situation/damage reports as per field units' observations and reports from the general public.
- Provide supplies, equipment, and personnel as requested.
- Provide buses for evacuations and temporary shelters.
- Provide bus transportation resources.
- Coordinate mobilization of emergency transportation services.
- Use transportation communication links to provide damage assessment information.

ff. Metro Humane Shelter

- Perform functions in the EOC or on-scene as assigned.
- Provide EMA and/or EOC initial situation/damage reports as per field units' observations and reports from the general public.
- Provide supplies, equipment, and personnel as requested.
- Provide assistance in the prevention, detection, and control of rabies.

gg. National Guard

- Perform functions in the EOC or on-scene as assigned.
- Provide EMA and/or EOC initial situation/damage reports as per field units' observations and reports from the general public.
- Provide supplies, equipment, and personnel as requested.
- Provide aircraft for search and rescue operations.

hh. Parks and Recreation

- Perform functions in the EOC or on-scene as assigned.
- Provide EMA and/or EOC initial situation/damage reports as per field units' observations and reports from the general public.
- Provide supplies, equipment, and personnel as requested.
- Provide facilities for emergency shelter, food, and water distribution points, child care facilities, as needed.
- Provide receiving and distribution sites.
- Assist with the delivery of donated goods.

ii. Personnel Board

- Perform functions in the EOC or on-scene as assigned.
- Provide EMA and/or EOC initial situation/damage reports as per field units' observations and reports from the general public.
- Provide supplies, equipment, and personnel as requested.
- Provide for the recruitment of manpower needs by the city for the organization and operation of the congregate care facilities (shelter/mass feeding).
- Coordinate employee issues with Finance.
- Provide emergency hire program.

jj. Public (General)

- Provide observations and reports to EMA and/or EOC.
- Provide manpower, supplies, and equipment.

kk. Purchasing Department

- Perform functions in the EOC or on-scene as assigned.
- Provide EMA and/or EOC initial situation/damage reports as per field units' observations and reports from the general public.
- Provide supplies, equipment, and personnel as requested.
- Establish a system for the coordination and acquisition of supplies, equipment, and services in support of emergency response efforts.

ll. Public Works

- Perform functions in the EOC or on-scene as assigned.
- Provide EMA and/or EOC initial situation/damage reports as per field units' observations and reports from the general public.
- Provide supplies, equipment, and personnel as requested.
- Assist with initial infrastructure damage assessment of horizontal construction, i.e., roads, bridges, storm sewers, dams, etc.
- Provide heavy equipment to support rescue operations.
- Provide technical information on damaged structures.
- Provide traffic control signs and barricades, and operational control of traffic signals.
- Assist with the identification of evacuation routes and keep evacuation routes clear of stalled vehicles.
- Coordinate the disposal of solid waste from congregate care facilities (shelter/mass feeding).
- Coordinate emergency utility support requirements with public and private utilities.

mm. Radio Amateur Civil Emergency Services (RACES)

- Perform functions in the EOC or on-scene as assigned.
- Provide EMA and/or EOC initial situation/damage reports as per field units' observations and reports from the general public.
- Provide supplies, equipment, and personnel as requested.
- Assist with the warning and emergency information dissemination.
- Provide communications support.
- Within capabilities, provide emergency radio communication links between the CAT/EOC and shelters.

nn. American Red Cross

- Perform functions in the EOC or on-scene as assigned.
- Provide EMA and/or EOC initial situation/damage reports as per field units' observations and reports from the general public.
- Provide supplies, equipment, and personnel as requested.
- Provide and deploy damage assessment teams to augment local damage assessment.
- Assist with emergency information dissemination.
- Provide supplementary medical and nursing care in American Red Cross shelters and other public health services upon request, and within their capabilities.
- Provide mass care for major fire scenes.
- Provide fire aid centers for noncritical injuries.
- Assist with coordination of needed blood, blood products, and vaccines.
- Coordinate with Jefferson County EMA and other organizations to establish and provide shelter reception centers.
- Establish and coordinate mass feeding.
- Provide mobile canteen service to victims and emergency service workers.
- Provide food, clothing, housing, household furnishings, medical, bedding and linens, occupational supplies, and other necessities to disaster victims.

oo. Risk Management

- Perform functions in the EOC or on-scene as assigned.
- Provide EMA and/or EOC initial situation/damage reports as per field units' observations and reports from the general public.
- Provide supplies, equipment, and personnel as requested.
- Coordinate, administer county insurance programs.

pp. Salvation Army

- Provide supplies, equipment, and personnel as requested.
- Provide mass care for major fire scenes.
- Provided fixed and mobile feeding sites.
- Provide various emergency services to include case work services, financial counseling, and a wide variety of emergency aid to people in need (e.g., food boxes, clothing, bedding, cash grants for emergency lodging, cleanup kits and many other specific assistance needs).
- Provide counseling to disaster victims.

qq. Schools (Facilities)

- Provide supplies, equipment, and personnel as requested.

rr. Schools (Districts)

- Provide supplies, equipment, and personnel as requested.
- Provide buses in support of evacuation and shelters.
- Provide school facilities for shelter and feeding.

ss. Specialists (When Necessary)

- Perform functions in the EOC or on-scene as assigned.

tt. Tax Assessor

- Provide appraisers to assist with damage assessment.

uu. Utilities

- Conduct infrastructure damage assessment of utility "life lines" (water, power, natural gas, telecommunications, sewer, waste services) owned by each utility.
- Provide supplies, equipment, and personnel as requested.
- Provide EOC management oversight of utility actions to ensure that the needs of special populations and individuals are provided for.

vv. Veterinarians

- Provide assistance in the prevention, detection and control of rabies.

ww. Volunteer Organizations

- Perform functions in the EOC or on-scene as assigned.
- Provide EMA and/or EOC initial situation/damage reports as per field units observations and reports from the general public.
- Provide supplies, equipment, and personnel as requested.
- Respond to search and rescue missions as requested and within their capabilities.
- Assist with meeting the needs of special populations and individuals.
- Assist in response and recovery involving donated goods and services.

4. State of Alabama

a. If local capabilities are exceeded, and a local emergency has been declared, State government agencies can augment assistance to local government to meet the emergency needs of victims during declared emergencies/disasters. Requests for state assistance are processed through the Jefferson County EMS.

b. The Alabama Emergency Management Agency (AEMA) receives and coordinates requests for state assistance. The Governor may declare a "state of emergency" to authorize use of state resources.

---

5. Federal Government

The Federal Response Plan (FRP) facilitates the provision of Federal assistance to state and local governments during major disasters. The FRP uses a functional approach to group the type of Federal assistance which State/local government is most likely to need under Emergency Support Functions (ESF). Each ESF is headed by a primary federal agency which has been selected based on its authorities, resources, and capabilities in the particular functional area.

Jefferson County, Alabama
Comprehensive Emergency Management Plan

Section 1: Basic Plan

III. Concept of Operations

A. General

1. The Emergency Management Council is responsible for the direction, control, and coordination of emergency management activities in Jefferson County.

2. The primary objective for emergency management in Jefferson County is to provide a coordinated effort from all supporting county and city departments in the preparation for, response to, and relief from injury, damage and suffering resulting from either a localized or widespread disaster. The EMA Coordinator is the focal point for emergency management activities within the county. However, emergency management responsibilities extend beyond this office, to all city/county government departments/agencies, and ultimately, to each individual citizen.

3. It is important to note that a basic responsibility for emergency planning and response also lies with individuals and heads of households. When the situation exceeds the capabilities of individuals, families, and volunteer organizations, a city/county emergency may exist. It is then the responsibility of government to undertake the effects of disasters. Local government has the primary responsibility for emergency management activities. When the emergency exceeds local government capability to respond, the EMA Coordinator will request assistance from mutual aid counties and/or the State government; the Federal Government will provide assistance to the State when requested, if possible. In addition, private sector and voluntary organizations may be requested to provide aid and assistance.

4. In addition to the EMA Coordinator, emergency management is the day-to-day function of certain city and county agencies, such as the Police and Fire Departments. While the routine functions of most city and county agencies are not of an emergency nature, pursuant to this plan, all officers and employees of the cities and county will plan to meet emergencies threatening life or property. This entails a day-to-day obligation to assess and report the impact of an emergency or disaster event. It requires monitoring conditions and analyzing information that could signal the onset of one of these events. Disasters will require city and county departments to perform extraordinary functions. In these situations, every attempt will be made to preserve organizational integrity and assign tasks which parallel the norm. However, it may be necessary to draw on people's basic capacities and use them in areas of greatest need. Day-to-day functions that do not contribute directly to the emergency operation may be suspended for the duration of any emergency. Efforts that would normally be required to perform those functions may be redirected to accomplish emergency tasks.

---

5. This plan does not contain a listing of resources. The EMA Coordinator will ensure that a resource inventory including source and quantity is kept current. The resource list will be maintained in the EOC. The EMA Coordinator should also be familiar with resources available from local private sector and volunteer organizations as well as from State government. Unique resources which may not be available locally (i.e., radiological and chemical analysis, environmental assessment, biological sampling, contamination survey, etc.) should be requested through the AEMA.

B. Emergency Management Phases

The county will meet its responsibility for protecting life and property from the effects of hazardous events by acting within each of the four phases of emergency management.

1. **Mitigation.** Actions accomplished before an event to prevent it from causing a disaster, or to reduce its effects if it does, save the most lives, prevent the most damage and are the most cost effective. See also Section 2, Mitigation of this plan. County and city departments will enforce all public safety mandates including land use management and building codes; and recommend to governing bodies legislation required to improve the emergency readiness of the county.

2. **Preparedness.** Preparedness consists of almost any predisaster action which is assured to improve the safety or effectiveness of disaster response. Preparedness consists of those activities that have the potential to save lives, lessen property damage, and increase individual and community control over the subsequent disaster response. Departments/agencies within the county will remain vigilant to crises within their areas of responsibility. All departments/agencies shall prepare for disasters by developing detailed SOPs to accomplish the extraordinary tasks necessary to integrate the department/agency's total capabilities into a city/county disaster by response. Disaster SOPs must complement this plan. Departments/agencies shall ensure that their employees are trained to implement emergency and disaster procedures and instructions. Departments/agencies shall validate their level of emergency readiness through internal drills and participation in exercise selected by the EMA Coordinator. Other government jurisdictions within and outside city/county boundaries shall also be encouraged to participate in these exercises. Exercise results shall be documented and used in continual planning effort to improve the county's emergency readiness posture. This joint, continuous planning endeavor shall culminate in revisions to this plan in the constant attempt to achieve a higher state of readiness for an emergency or disaster response.

3. **Response.** The active use of resources to address the immediate and short-term effects of an emergency or disaster constitutes the response phase and is the focus of department/agency emergency and disaster Standard Operating Procedures, mutual-aid agreements, and this plan. Emergency and disaster incident responses are designed to minimize suffering, loss of life, and property damage, and to speed recovery. They include initial damage assessment, emergency and short-term medical care, and the return of vital life-support systems to minimum operating conditions. When any department/agency within the county receives information about a potential emergency or disaster, it will conduct an initial assessment, determine the need to alert others, and set in motion appropriate actions to reduce risk and potential impacts. Emergency response activities will be as described in department/agency SOPs and may involve activating the Emergency Operations Center (EOC) for coordination of support. Departments/agencies will strive to provide support to warning and emergency public information, save lives, and property, supply basic human needs, maintain or restore essential services, and protect vital resources and the environment. Responses to declared emergencies and disasters will be guided by this plan.

4. **Recovery.** Emergency and disaster recovery efforts aim at returning to pre-disaster community life. They involve detailed damage assessments, complete restoration of vital life-support systems, financial assistance, and long-term medical care. There is no definite point at which response ends and recovery begins. However, generally speaking, most recovery efforts will occur after the emergency organization is deactivated and departments/agencies have returned to predisaster operation, and will be integrated with day-to-day functions.

C. Functional Annex Concept

1. "Functional Annexes" represent groupings of types of assistance activities that citizens are likely to need in times of emergency or disaster. In Alabama, county and state Comprehensive Emergency Management Plans, are organized by related emergency functions. The Federal Response Plan is similarly organized into "emergency support functions (ESFs)." During emergencies within Jefferson County, the EMA coordinator will determine which functional annexes or ESFs are activated to meet the disaster response needs. See "Section 4, Response-Functional Annexes" for further details. The Federal Government, through the State EOC will respond to Jefferson county requests for assistance through the Federal ESF structure.

2. Within the EOC, requests for assistance will be tasked to the particular "emergency function" or "EOC Branch or Unit" for completion. A lead agency for each emergency function is indicated, and will be responsible for coordinating the delivery of that assistance to the emergency area. The lead agency will be responsible for identifying the resources that will accomplish the mission, and will coordinate the resource delivery.

Jefferson County, Alabama
Comprehensive Emergency Management Plan

Section 1: Basic Plan

IV. Administration and Logistics

A. General

During and after emergency/disaster events normal fiscal and administrative functions and regulations may need to be temporarily modified or suspended in order to support emergency operations in a timely manner. Additionally, if certain emergency costs can be documented, certain reimbursements from State and Federal sources may be possible.

B. Policies

It is the policy of the Jefferson County Emergency Management Council that:

1. All departments/agencies shall assure the safety of cash, checks, accounts receivable, and assist in the protection of other valuable documents/records.

2. All departments/agencies shall designate personnel to be responsible for documentation of disaster operations and expenditures. Emergency expenditures will be incurred in accordance with existing jurisdictional emergency purchasing procedures.

3. During the emergency operations, nonessential administrative activities may be suspended, and personnel not assigned to essential duties may be assigned to other departments to provide emergency support.

4. Each department/agency shall keep an updated inventory of its personnel, facilities, and equipment resources as part of their SOPs.

C. Administration

1. During an emergency or disaster, administrative procedures may have to be suspended, relaxed, or made optional in the interest of protecting life or property. Departments/agencies are authorized to take necessary and prudent actions in response to disaster/emergency incidents.

2. Normal procedures which do not interfere with timely accomplishment of emergency tasks, will continue to be used. Those emergency administrative procedures which depart from "business-as-usual" will be described in detail in department/agency SOPs.

3. Departments/agencies are responsible for keeping records of the name, arrival time, duration of utilization, departure time, and other information relative to the service of emergency workers, as well as documentation of the injuries, lost or damaged equipment, and any extraordinary costs.

D. Fiscal

1. Local government purchasing personnel shall facilitate the acquisition of all supplies, equipment, and services necessary to support the emergency response actions of departments/agencies.

2. A complete and accurate record of all purchases, a complete record of all properties commandeered to save lives and property, and an inventory of all supplies and equipment purchased in support of the emergency response shall be maintained.

3. Though certain formal procedures may be waived, this in no way lessens the requirement for sound financial management and accountability. Departments/agencies will identify personnel to be responsible for documentation of disaster costs and utilize existing administrative methods to keep accurate records separating disaster operational expenditures from day-to-day expenditures. Documentation will include: logs, formal records, and file copies of all expenditures, receipts, personnel time sheets.

4. A separate Emergency Operations Center (EOC) "Finance Section" may be formed to handle the monetary and financial functions during large emergencies, disasters. See the Jefferson County EMA EOC Position Checklists Manual for details.

E. Logistics

1. Departments/agencies responding to emergencies and disasters will first use their available resources. When this plan is implemented, the EOC Logistics Section becomes the focal point of procurement, distribution and replacement of personnel, equipment, and supplies. The Logistics Section will also provide services and equipment maintenance beyond the integral capabilities of elements of the emergency organization. Scarce resources will be allocated according to established priorities and objectives of the EOC.

2. Logistics will be needed to support the field operations, the EOC operations, and disaster victims.

3. All departments/agencies are expected to maintain an inventory of all nonconsumable items, to include their disposition after the conclusion of the emergency proclamation. Items that are not accounted for, or that are placed in local government inventory as an asset will not be eligible for reimbursement.

F. Insurance

Local governments, agencies shall maintain insurance for property, workers' compensation, general and automotive liability. Insurance coverage information will be required by the Federal Government in the post disaster phase as per 44 CFR "subpart 1." Information on insurance needs to be available following a disaster.

Further, all local jurisdictions and departments are responsible to maintain adequate levels of insurance.

Jefferson County, Alabama
Comprehensive Emergency Management Plan

Section 1: Basic Plan

V. Direction and Control

A. General

1. The chief executive of the local government in the jurisdiction in which the emergency occurs will exercise direction and control activities within that jurisdiction. He/she will coordinate with the EMA Coordinator who is responsible for implementing this Plan or portions of this Plan. Each jurisdiction shall establish Standard Operating Procedures (SOPs) to control and direct response actions. The EMA Coordinator will coordinate actions between local governments and agencies as necessary, and direct response actions in unincorporated areas of the county. In cases where local resources to contend with an emergency do not exist or have been depleted, the affected chief executive, in coordination with the EMA Coordinator, should request state aid through the state EOC.

2. The chief executive of the local government may declare a "State of Emergency" to expedite access to local resources needed to cope with the incident. If the needed response exceeds these local capabilities, a disaster has occurred. The chief executive may, by emergency proclamation, use local resources and employees as necessary, and alter functions of departments and personnel, as necessary. The Jefferson County Emergency Management Council has the authority to declare a "state of emergency" in support of a local government emergency. If the situation is beyond local capability, a request for State and/or Federal assistance may be in the original proclamation, or included in a second proclamation presented to the Governor through the Alabama Emergency Management Agency. If State and/or Federal resources are made available, they will be under the operational control of the EMA Coordinator/EOC.

3. On behalf of the Emergency Council, the EMA Coordinator has the responsibility for coordinating the entire emergency management program. The Coordinator makes all routine decisions and advises the officials on courses of action available for major decisions. During emergency operations the Coordinator is responsible for the proper functioning of the EOC. The Coordinator also acts as a liaison with State and Federal emergency agencies, and neighboring counties.

4. The Emergency Operations Center (EOC) is the central point for emergency management operations. The purpose of this central point is to ensure harmonious response when the emergency involves more than one political entity and several response agencies. Coordination and supervision of all services will be through the EOC section chiefs and the EMA Coordinator to provide for the most efficient management of resources.

---

5. During emergency situations, certain agencies will be required to relocate their center of control to the EOC. During large scale emergencies, the EOC will become the seat of government during the duration of the crisis. However, in some situations it may be appropriate for some agencies to operate from an alternate site other than the EOC or their primary location.

6. Specific persons and agencies are responsible for fulfilling their obligations as presented in the Basic Plan and individual annexes. Department/agency heads will retain control over their employees and equipment. Each department/agency shall develop Standard Operating Procedures (SOPs) to be followed during response operations.

7. Department/agency heads and other officials legally administering from their office may perform their emergency functions(s) on their own initiative if, in their judgment, the safety or welfare of citizens of the county are threatened. The EMA Coordinator should be notified as rapidly as possible.

8. During an EOC activation, the appropriate emergency services will be represented in the EOC and will coordinate their activities under the supervision of the EMA Coordinator. EOC procedures are described in Annex 1.

9. Additional information on Direction and Control is found in Annex 1.

B. Crisis Monitoring

1. The Jefferson County EMA is the county's 24 hour "crisis monitor." The EMA Coordinator provides an ongoing independent analysis of incoming information. As emergency situations threaten or occur, the EMA Coordinator may convene a "Crisis Action Team (CAT)" to facilitate the process of evaluation and incident planning, and possible activation and implementation of emergency functions and resources. The CAT will also be used to support "Incident Commanders" in field situations.

2. The CAT is a flexible, supporting/coordinating service that could be one person at home facilitating the coordination of personnel and resources to an incident scene or several people convening in the EOC or on-scene to assist the "Incident Commander" as needed.

3. Core members of the EMA CAT include Emergency Management Council Chairperson, EMA Coordinator, Fire Services Coordinator, Law Enforcement Coordinator, Public Works Coordinator, Health Services Coordinator, Legal Services Coordinator, and other services coordinators as required. However, any department (division)/agency could be called upon to provide a representative to the CAT. Exactly who is called and ultimately how many people will serve on the CAT is dependent upon the situation and the emergency functions that will be activated.

4. A CAT may be activated in support of an emergency situation within a particular local government. Normal local government CAT membership includes the Mayor, County EMA Coordinator, Fire Chief, Police Chief, Public Works Director, City Attorney, County Health Officer as needed, and any other municipal department as required.

C. Levels of Emergency

1. Localized Emergency. The principal of graduated response will be used in responding to localized disasters defined as an incident within a local government. The initial response will be from emergency personnel dispatched by normal procedures. Their assessment of the situation will determine if additional resources are needed. Departments/agencies may be called upon to provide additional resources. Mutual aid and the local government CAT Team/EOC may provide additional support if resource needs are beyond existing city capability. Activation of the Jefferson County EOC may not be necessary during a "localized" emergency.

2. Widespread Disaster. Hurricanes, tornadoes, floods, snow/ice storms are considered the most probable widespread disaster which could impact the entire county and adjacent areas. It is anticipated that a full activation of the EOC will be required to coordinate the county's response.

3. Graduated Response. Most disasters will require a graduated response involving only those persons necessary to handle the situation. For this purpose, four levels of response will be used:

   a. **Level One Emergency.** A "level one" emergency is a common emergency situation that occurs on a frequent basis. The responsibility for control of the incident rests with the responding department.

   b. **Level Two Emergency.** Should an incident remain unresolved, the emergency status will rise to a "level two" emergency. Level two incidents involve routine assistance from internal and/or external agencies including mutual aid. Command and control is still the responsibility of the primary response department. Notification of the EMA Coordinator and Emergency Council is necessary. A CAT Team may be activated and/or partial activation of the EOC may occur. The EMA Coordinator or designee may go to the Incident Command Post to facilitate response coordination. The State EOC is notified, and state assistance may be requested.

   c. **Level Three Emergency.** Should the incident begin or escalate to a situation where nonroutine assistance is required or anticipated, a "level three" emergency will be declared. The EOC will activate at this level. A level three could be a major single site event or a county-wide event. This level of emergency will be used for all natural, manmade or major technological disasters. The State EOC is notified, communication and coordination is maintained. State assistance may be requested.

   d. **Level Four Emergency.** This level of emergency is used for "catastrophic" state-wide or regional events. The Jefferson County EOC is fully activated. State and Federal assistance will be requested and is required.

D. On-Scene Management/ICS

1. On-scene response to emergencies follows the concept of the Incident Command System (ICS).

2. The person in charge at the incident is the on-scene Incident Commander who is responsible for ensuring each agency on scene can carry out its responsibilities.

3. Upon arriving at an incident scene, the Incident Commander should:

   - Assess the situation and identify hazards.
   - Develop objectives (tasks to be done).
   - Ensure appropriate safety and personnel protective measures.
   - Develop an action plan and priorities.
   - In coordination with the EOC, contact appropriate agencies or personnel with expertise and capability to carry out the incident action plan.
   - Coordinate, as appropriate, with other first responders.

4. When more than one agency is involved at an incident scene, the Incident Command Agency and other responding agencies should work together to ensure that each agency's objectives are identified and coordinated.

5. Team problem solving should facilitate effective response. Other agency personnel working in support of the Incident Command Agency will maintain their normal chain of command, but will be under control of the on-scene Incident Commander.

6. The on-scene Incident Commander may designate an Information Officer to work with the news media at an incident. This may include coordinating agency media releases and arranging contacts between the media and response agencies. If additional support is needed, a Crisis Action Team (CAT) and/or the EOC may be activated.

E. EOC Activation and Staffing

1. The Emergency Operations Center (EOC) is the key to successful response and recovery operations. With decision and policymakers located together, personnel and resources can be used efficiently. Coordination of activities will ensure that all tasks are accomplished, minimizing duplication of efforts.

2. Depending upon the severity and magnitude of the disaster, activation of the EOC may not be necessary, may only be partially required, or may require full activation. Partial activation would be dictated by the characteristics of the disaster and would involve only those persons needing to interact in providing the coordinated response.

3. The EOC may be fully activated by decision of the Crisis Action Team (CAT) or the EMA Coordinator. When the decision is made to activate the EOC, the EMA Coordinator will notify the appropriate EOC organization staff members to report to the EOC. The EOC Management staff will take further action to notify and mobilize the appropriate organizations and dispatch centers which they are responsible for coordinating.

4. EOC activation levels will generally follow the "emergency levels" as discussed in Section C above.

5. EOC Operations and Staffing. (See Annex 1, Managing Emergency Operations, for further information.) Complete details, job descriptions, and checklist of tasks are contained in the Emergency Operations Center (EOC) Position Checklists Manual, published separately.

F. Controls, Continuity of Operations

1. In an emergency there will be two levels of control. The first level of control will be at the scene of the incident. The second level will be at the EOC where overall coordination will be exercised.

2. In a single site emergency, the governing body having jurisdiction will respond to the scene. The on-scene management will fall under the jurisdiction of the local department best qualified to conduct the rescue, recovery, and control operations. The department's senior representative at the scene will become the on-scene commander and will be responsible for the overall recovery operations. The field Incident Commanders are local officials, usually fire or police officers. The local coordination and commitment authority for local resources is retained by the local elected officials, and delegated as appropriate.

3. During wide-spread emergencies, decision making authority and control of the emergency is retained by the Emergency Management Council through the actions of CAT Team or activation of the county EOC.

4. The county EOC, once activated, directs and controls a response to an emergency or disaster. It is organized and will function according to the National Interagency Incident Management System's (NIIMS) Incident Command System (ICS) principles.

5. EOC Incident Command (EOC Incident Manager position) will normally be vested in the department deciding to activate the EOC to support their emergency response activities, unless otherwise specified by the Emergency Management Council, who will be notified and briefed by the EMA Coordinator as soon as possible. Consistent with the modular component of National Interagency Incident Management System's (NIIMS) Incident Command System (ICS), the EOC may be partially activated to coordinate support for an on-scene incident commander, without activating the full emergency organization.

6. During the effective period of any declared emergency, the EMA Coordinator through the EOC Incident Manager directs and controls all emergency response activities and employs all necessary emergency resources according to the provisions of this plan.

7. To ensure a line of succession, each department/agency is directed to assign 3 or more alternates for each key emergency position. Lines of succession shall be provided to the EMA Coordinator.

G. Facilities

1. Emergency Operations Center

   a. The county EOC is located underneath the parking deck adjacent to Birmingham City Hall.

   b. Each jurisdiction is encouraged to establish an on-scene command post. These facilities will link to the county EOC via radio or telephone.

2. Department/Agency Operating Locations

   a. Each department/agency is directed to establish a primary location and alternate location from which to establish direction and control of its respective activities in an emergency or disaster. This may be from the EOC, or other location, depending upon the circumstances.

3. Communications (Also see Annex 5, Communication Systems)

   a. Most departments/agencies involved in disaster operations will maintain operations or dispatch centers that will control the operations of the emergency forces under their control.

   b. Any department/agency operating from another location will maintain contact with the EOC through direct redundant communication, such as telephone and radio.

   c. All departments/agencies are responsible to ensure that communication systems are in place between EOC representatives and their department/agency.

   d. Each department/agency will bring to the EOC their own portable radio, charger, spare batteries, and cellular telephones. Additional communications equipment will be provided at the EOC.

H. Military Support

Military support to Jefferson County will be requested through the AEMA. Once assigned, resources shall be coordinated by the county EOC.

I. Continuity of Government

1. Succession of Authority (Alabama Code 29-3-15/16)

   a. A community's ability to respond to an emergency must not be restrained by the absence of an elected official or key department head. Therefore, to ensure continuity of government, each local government in the county will develop a Continuity of Government Succession List. This list will name who will be the decision maker if an elected official or department head is not available. At least two people should be listed and prioritized for each key position.

b. The line of succession for the County Commission is from the President to the members of the Commission in order of their seniority on the Commission.

c. The City Councils will determine the line of succession to the Mayors.

d. The senior EMA officer will succeed the EMA Coordinator followed by officers in order of their seniority.

e. Lines of succession to each department head will be determined by the appropriate county or city governing body or by the department's Standard Operating Procedures.

2. Preservation of Records

All departments/agencies will develop SOPs to guarantee the preservation of vital public records, to include their reconstitution if necessary, during and after emergencies. In general, vital public records include those considered absolutely essential to the continued operation of local government; considered absolutely essential to the local government's ability to fulfill its responsibilities to the public; required to protect the rights of individuals and the local government; and essential to restoration of life support services. Documentation of actions taken during an emergency or disaster is a legal requirement.

J. Plan Maintenance

1. If a plan is to be effective, its contents must be known and understood by those who are responsible for its implementation. The EMA Coordinator will brief the appropriate officials concerning their roles in emergency management and this plan in particular.

2. All agencies will be responsible for developing and maintaining their respective segments of the plan. The EMA Coordinator will be responsible for ensuring all officials involved in this plan conduct an annual review of the plan.

3. The EMA ensures that necessary changes and revisions to the plan are prepared, coordinated, published and distributed. The plan will undergo revision whenever:

   - It fails during emergency.
   - Exercises, drills reveal deficiencies or "shortfalls."
   - City or county government structure changes.
   - Community situations change.
   - State requirements change.
   - Any other condition occurs that causes conditions to change.

4. EMA will maintain a list of individuals and organizations which have controlled copies of the plan. Only those with controlled copies will automatically be provided updates and revisions. Plan holders are expected to post and record these changes. Revised copies will be dated and marked to show where changes have been made.

5. The plan shall be activated at least once a year in the form of a simulated emergency to provide practical controlled operational experience to those individuals who have EOC responsibilities. Response to radiological and hazardous materials incidents must be exercised at least once a year.

6. The Title III, Local Emergency Planning Committee, as a state Advisory Agency, will review applicable portions of this plan.

Jefferson County, Alabama
Comprehensive Emergency Management Plan

Section 1: Basic Plan

VI. Attachments

This section of the Basic Plan contains the following information:

A. Primary/Support Matrix

B. Job Aids

C. Acronym List

A. Primary Matrix

Agencies, Departments	Managing Emergency Operations (1)	Situation Reporting (2)	Damage Assessment (3)	Alert, Warning, Notification (4)	Emergency Public Information (5)	Communication Systems (6)	Resource Management (7)	Human Resources (8)	Search & Rescue (9)	Public Works (10)	Public Health Services (11)	Animal Considerations (12)	Fire Services (13)	Emergency Medical Services (14)	Law Enforcement Services (15)	Coroner/Medical Examiner (16)	Population Relocation (17)	Transportation (18)	Human Services (19)	Donated Goods & Services (20)	Emergency Fiscal & Administrative (21)
Ambulance Service	S	S	S		S	S	S							S		S		S			S
American Red Cross	S	S	S	S	S	S	S	S			S		S	S				S	S	P	S
Birmingham Communications Dept.	S	S	S		S	S	S														S
Birmingham Equipment Management	S	S	S		S	S	S											S			S
BREMSS	S	S	S		S	S	S							S							S
Building Inspection Services	S	S	S		S	S	S			S											S
Business & Industry	S	S	S		S	S	S				S			S	S	S			S		S
Campuses, Universities	S	S	S		S	S	S												S		S
Churches	S	S	S		S	S	S												S		S
Civil Air Patrol	S	S	S		S	S	S		S												S
Community Service Organizations	S	S	S		S	S	S												S		S
Data Processing	S	S	S		S	S	S														S
Department of Human Resources	S	S	S		S	S	S											S	S	P	S
Emergency Management Council	P	S	S		S	S	S														S
Finance	S	S	S		S	S	S														S
Fire Services	S	S	S	S	S	S	S		S		S		P	P		S	S	S			S
Funeral Directors; Associations	S	S	S		S	S	S									S					S
Greater Birmingham Humane Society	S	S	S		S	S	S				S	P									S
Hospitals	S	S	S		S	S	S				S			S							S
Jefferson County Communications	S	S	S		S	S	S														S
Jefferson County Coroner/Medical Examiner's Office	S	S	S		S	S	S										P				S
Jefferson County Department of Health	S	S	S		S	S	S				P								S		S
Jefferson County Emergency Management Department	P	P	P	P	P	P	P	S	P	S	S	S	S	S	S	S	S	P	P	S	P

A. Support Matrix

Agencies, Departments	1 Managing Emergency Operations	2 Situation Reporting	3 Damage Assessment	4 Alert, Warning, Notification	5 Emergency Public Information	6 Communication Systems	7 Resource Management	8 Human Resources	9 Search & Rescue	10 Public Works	11 Public Health Services	12 Animal Considerations	13 Fire Services	14 Emergency Medical Services	15 Law Enforcement Services	16 Coroner/Medical Examiner	17 Population Relocation	18 Transportation	19 Human Services	20 Donated Goods & Services	21 Emergency Fiscal & Administrative
Jefferson County Fleet Services	S	S	S		S	S	S												S		S
Jefferson County General Services	S	S	S		S	S	S													S	S
Law Enforcement	S	S	S	S	S	S	S		P		S	S	S		P	S	P	S			S
Management Information Services	S	S	S		S	S	S							S					S		S
Media	S	S	S	S	S	S	S														S
Metro Area Express	S	S	S		S	S	S								S	P	S				S
Metro Humane Shelter (Note: Name will Change)	S	S	S		S	S	S				S	P									S
National Guard	S	S	S		S	S	S		S												S
Parks & Recreation	S	S	S		S	S	S	P											S	S	S
Personnel Board	S	S	S		S	S	S												S	S	S
Public (General)	S	S	S		S	S	S														S
Purchasing Department	S	S	S		S	S	S														S
Public Works	S	S	S		S	S	S		S	P			S		S		S	S	S		S
RACES	S	S	S	S	S	S	S		S										S		
Risk Management	S	S	S		S	S	S														S
Salvation Army	S	S	S		S	S	S	S					S						S		S
Schools (Districts)	S	S	S	S	S	S	S										S	P	S		S
Tax Assessor	S	S	S		S	S	S														S
Utilities	S	S	S		S	S	S			S									S		S
Veterinarians	S	S	S		S	S	S				S	S							S		S
Volunteer Organizations	S	S	S		S	S	S		S										S	S	S

Appendix B: Job Aids

Job Aid 2.1: Planning Team Participants

Individuals/Organizations	What They Bring to the Planning Team
Senior Official (elected or appointed) or designee	• Support for the homeland security planning process. • Government intent by identifying planning goals and essential tasks. • Policy guidance and decision-making capability. • Authority to commit the jurisdiction's resources.
Emergency Manager or designee	• Knowledge about all-hazard planning techniques. • Knowledge about the interaction of the tactical, operational, and strategic response levels. • Knowledge about the prevention, protection, mitigation, response, and recovery strategies for the jurisdiction. • Knowledge about existing mitigation, emergency, continuity, and recovery plans.
EMS Director or designee	• Knowledge about emergency medical treatment requirements for a variety of situations • Knowledge about treatment facility capabilities • Specialized personnel and equipment resources • Knowledge about how EMS interacts with the Emergency Operations Center and incident command
Fire Chief or designee	• Knowledge about fire department procedures, on-scene safety requirements, hazardous materials response requirements, and search-and-rescue techniques. • Knowledge about the jurisdiction's fire-related risks. • Specialized personnel and equipment resources.
Police Chief or designee	• Knowledge about police department procedures; on-scene safety requirements; local laws and ordinances; explosive ordnance disposal methods; and specialized response requirements, such as perimeter control and evacuation procedures. • Knowledge about the prevention and protection strategies for the jurisdiction. • Knowledge about fusion centers and intelligence and security strategies for the jurisdiction. • Specialized personnel and equipment resources.
Public Works Director or designee	• Knowledge about the jurisdiction's road and utility infrastructure. • Specialized personnel and equipment resources.

Job Aid 2.1: Planning Team Participants (Continued)

Individuals/Organizations	What They Bring to the Planning Team
Public Health Officer or designee	• Records of morbidity and mortality. • Knowledge about the jurisdiction's surge capacity. • Understanding of the special medical needs of the community. • Knowledge about historic infectious disease and syndromic surveillance. • Knowledge about infectious disease sampling procedures.
Hazardous Materials Coordinator	• Knowledge about hazardous materials that are produced, stored, or transported in or through the community. • Knowledge about U.S. Environmental Protection Agency (EPA), Occupational Safety and Health Administration (OSHA), and U.S. Department of Transportation (DOT) requirements for producing, storing, and transporting hazardous materials and responding to hazardous materials incidents.
Hazard Mitigation Specialist	• Knowledge about all-hazard planning techniques. • Knowledge of current and proposed mitigation strategies. • Knowledge of available mitigation funding. • Knowledge of existing mitigation plans.
Transportation Director or designee	• Knowledge about the jurisdiction's road infrastructure. • Knowledge about the area's transportation resources. • Familiarity with the key local transportation providers. • Specialized personnel resources.
Agriculture Extension Service	• Knowledge about the area's agricultural sector and associated risks (e.g., fertilizer storage, hay and grain storage, fertilizer and/or excrement runoff).
School Superintendent or designee	• Knowledge about school facilities. • Knowledge about the hazards that directly affect schools. • Specialized personnel and equipment resources (e.g., buses).
Social services agency representatives	• Knowledge about special-needs populations

Job Aid 2.1: Planning Team Participants (Continued)

Individuals/Organizations	What They Bring to the Planning Team
Local Federal asset representatives	• Knowledge about specialized personnel and equipment resources that could be used in an emergency. • Facility security and response plans (to be integrated with the jurisdiction's EOP). • Knowledge about potential threats to or hazards at Federal facilities (e.g., research laboratories, military installations).
NGOs (includes members of National VOAD [Voluntary Organizations Active in Disaster]) and other private, not-for-profit, faith-based, and community organizations	• Knowledge about specialized resources that can be brought to bear in an emergency. • Lists of shelters, feeding centers, and distribution centers. • Knowledge about special-needs populations.
Local business and industry representatives	• Knowledge about hazardous materials that are produced, stored, and/or transported in or through the community. • Facility response plans (to be integrated with the jurisdiction's EOP). • Knowledge about specialized facilities, personnel, and equipment resources that could be used in an emergency.
Amateur Radio Emergency Service (ARES)/Radio Amateur Civil Emergency Services (RACES) Coordinator	• List of ARES/RACES resources that can be used in an emergency.
Utility representatives	• Knowledge about utility infrastructures. • Knowledge about specialized personnel and equipment resources that could be used in an emergency.
Veterinarians/animal shelter representatives	• Knowledge about the special response needs for animals, including livestock.

Job Aid 3.1: Threat Analysis Worksheet

Threat Analysis Worksheet
Threat:
Potential Consequences: ☐ Catastrophic *(Mass fatalities/casualties, loss of governance and essential services, widespread damages)* ☐ Severe *(Numerous fatalities/ casualties, loss of essential services, and widespread damage)* ☐ Moderate *(Limited number of fatalities/casualties and damage to property)* ☐ Minor *(Little or no injuries and isolated damage)*

Probability of Occurring: ☐ High ☐ Medium ☐ Low	**Past History:** Has this type of incident occurred before? ☐ Yes ☐ No If yes, when? _____

Areas Likely to be Affected Most:
Probable Duration:
Potential Speed of Onset (Probable amount of warning time): • Minimal (or no) warning.   • 12 to 24 hours warning. • 6 to 12 hours warning.   • More than 24 hours warning.
Existing Warning Systems:
Does a Vulnerability Analysis Exist?* Yes ☐ No ☐

* *Developing a Vulnerability Analysis is Step 4 of the threat analysis process.*

Appendix C: Acronym List

Listed below are all of the acronyms that are used in this course.

CEO	Chief Elected Official
CONPLAN	U.S. Government Interagency Domestic Terrorism Concept of Operations
DHS	Department of Homeland Security
DOT	Department of Transportation
EMS	Emergency Medical Services
EOC	Emergency Operations Center
EOP	Emergency Operations Plan
EPA	Environmental Protection Agency
EPI	Emergency Public Information
FRERP	Federal Radiological Emergency Response Plan
FRP	Federal Response Plan
HSPD	Homeland Security Presidential Directive
ICS	Incident Command System
INRP	Initial National Response Plan
NIMS	National Incident Management Plan
NRP	National Response Plan
NWS	National Weather Service
OSHA	Occupational Safety and Health Administration
RACES	Radio Amateur Civil Emergency Services
SOP	Standard Operating Procedure(s)
US&R	Urban Search and Rescue
WEM	*Workshop in Emergency Management*

www.ingramcontent.com/pod-product-compliance
Lightning Source LLC
Chambersburg PA
CBHW070137290526
45789CB00002B/517